INDUSTRIAL ENGINEERING AND MANAGEMENT SCIENCE

Communications in Information Engineering

ISSN 2333-2115

Volume 2

PROCEEDINGS OF THE 2014 INTERNATIONAL CONFERENCE ON INDUSTRIAL ENGINEERING AND MANAGEMENT SCIENCE (IEMS 2014), 8–9 AUGUST 2014, HONG KONG

Industrial Engineering and Management Science

Editor

Garry Lee
Information Engineering Research Institute, Bellflower, CA, USA

CRC Press
Taylor & Francis Group
Boca Raton London New York Leiden

CRC Press is an imprint of the
Taylor & Francis Group, an **informa** business
A BALKEMA BOOK

Published by:
CRC Press/Balkema
P.O. Box 447, 2300 AK Leiden, The Netherlands
e-mail: Pub.NL@taylorandfrancis.com
www.crcpress.com – www.taylorandfrancis.com

ISBN 13: 978-1-138-02647-6 (hbk)
ISBN 13: 978-0-429-22724-0 (ebk)

Typeset by MPS Limited, Chennai, India

Industrial Engineering and Management Science – Lee (Ed)
© 2015 Taylor & Francis Group, London, ISBN 978-1-138-02647-6

Table of contents

Industrial Engineering and Management Science – Lee (Ed)
© 2015 Taylor & Francis Group, London, ISBN 978-1-138-02647-6

Preface

Information Management, Innovation Management and Industrial Engineering are becoming increasingly interesting to both academic researchers and management practitioners. It is essential to explore the enterprise management system from a theoretical viewpoint; it is also absolutely essential to the survival, growth and prosperity of any company to have some means to manage innovation in the process of Economic Globalization and under the Knowledge Economy environment.

The 2014 International Conference on Industrial Engineering and Management Science (IEMS 2014) provides two day's focus on the science and engineering that are the basis for management. The theme of the plenary session is "Information Management, Innovation Management and Industrial engineering", featuring invited speakers who will further explore this topic that is so significant for management. Concurrent sessions and a poster session will cover a wide range of topics and issues, including both contributed papers and special sessions developed on specific themes, all with a central focus on management. Topics will range from management theories such as experimental management to management practice methods and technology, such as logistics and supply chain management, knowledge management, information management, innovation management, future technology in service of regional industry, etc.

This book is the proceedings of the 2014 International Conference on Industrial Engineering and Management Science (IEMS 2014) held August 8–9, 2014 in Hong Kong. IEMS is published in the Communications in Information Engineering (CIE) series by CRC Press/Balkema, Taylor & Francis Group. This year, we received more than 100 papers; each paper has undergone a rigorous review process. Only high-quality papers are included and 23 papers have been published. IEMS 2014 focuses on three topics: (1) Information Technology, (2) Industrial Development and Industrial Engineering, and (3) Performance Evaluation. Some papers were highly evaluated by the scientific committees and the reviewers. We expect that the conference and its publication will be a trigger for further related research and technology improvements in this importance subject area.

On behalf of the Organizing Committee, we acknowledge IERI and CRC Press/Balkema, Taylor & Francis Group. We wish to express our heart-felt appreciation to the keynote and panel speakers. We gratefully acknowledge the support of the contributors to this volume and contributions for the proceedings were sought from all participants and all papers received were carefully refereed by peer reviewers.

Industrial Engineering and Management Science – Lee (Ed)
© 2015 Taylor & Francis Group, London, ISBN 978-1-138-02647-6

IEMS2014 Organizing Committee

KEYNOTE SPEAKER

Gerald Schaefer, Loughborough University, U.K.

GENERAL CHAIRS

Dehuai Yang, Huazhong University of Science and Technology, China
Minli Dai, Suzhou University, China

PUBLICATION CHAIRS

Garry Lee, Information Engineering Research Institute, USA
March Zhou, Hong Kong Education Society, Hong Kong

INTERNATIONAL COMMITTEE

Choi Sang Long, Universiti Teknologi Malaysia, Malaysia
Fei Hu, Guangdong University of Technology, China
Ying-Chieh Liu, Chang Gung University, Taiwan
Shuai Liu, Institute of Microelectronic CAD Hangzhou Electronic University, China
S. Nakanishi, Ritsumeikan University, Japan
A.L. Wang, Tongji University, China
Jui-Pin Yang, Shih-Chien University, Taiwan
M. Qiu, Fukuoka Institute of Technology, Fukuoka, Japan
DunGang Zang, Sichuan Agricultural University, China
Mardhiah Yaacob, Universiti Teknologi Malaysia, Malaysia

Section 1: Information technology

Industrial Engineering and Management Science – Lee (Ed)
© 2015 Taylor & Francis Group, London, ISBN 978-1-138-02647-6

Design identity system based on product identity and its case studies

Jun Hu
The College of Post and Telecommunication of WIT, China

Fei Hu
Guangdong University of Technology, China

ABSTRACT: This paper explores how to construct a brand identity by a system of design approaches and design thinking. The system includes product design identity, communication design identity, service design identity, mind design identity, strategy design identity and behaviour design identity, etc. Further, aimed at original brand manufacture, a cake model of design identity is refined from three levels of product individual, product series/cluster and product brand, and three parts of product identity, communication identity and mind identity. Some clues can be found to build brand identity from both multi-brand strategy and single-brand strategy, where mind identity is crucial and decisive. Finally, this paper rethinks the design identity system based on product identity in the new context.

1 INTRODUCTION

In the era of post financial crisis, Chinese manufactures are facing a more complicatedly global market. Their core competitiveness of low cost and low price cannot be kept on any more, so they are called for transforming and upgrading strongly and urgently. When facing the variable emotions and different behaviours of customers as well as the coessential technologies and similar surfaces of products, how to build a unique and effective brand identity by design activities remains to be a vital question.

Brand identity has already become an important part of design management instead of corporate identity. However, there are still not very explicit ways to achieve brand identity directly and availably. Hence, this paper tries to integrate all kinds of design approaches, and blend design thinking into system and strategy, so as to construct and enhance an original brand quickly and validly.

2 LITERATURE REVIEW

2.1 *About corporate identity*

According to identity researchers in the French school (Larçon and Reitter, 1979; Moingeon, 1991; Moingeon and Ramanantsoa, 1995; Reitter and Ramanantsoa, 1985; Van Riel, 1997), corporate identity was the set, system or combination of interdependent characteristics of the organization that gave it specificity, stability and coherence, and thus made it identifiable. However, as Balmer and Soenen (1998) mentioned, it turned into a dissimilation, that is, practitioners tended to focus on visual identity and neglected other factors excessively, so that corporate identity was regarded as just the arrangement and application of a logo, which was superficial and misapprehensive. Jie Chen (2005) built a new CIS model that was product identity supported, and tried to communicate with external stakeholders via brand, which contained product identity, service identity, communication identity and strategy identity, behaviour identity and mind identity. However, there were not clear and explicit relationships, paths and methods in this paper.

2.2 *About brand identity*

Whether as belief, experience or symbol, a brand is a distinguishing feature to separate a company's supply of goods and services from competitors. Many researchers focussed on brand identity and brand image[1] (Abratt, 1989), whereas Dutton and Dukerich (1991) focussed on the relationship between brand identity and its perceived external image. Balmer (1998) defined that brand identity was to articulate the corporate ethos, aims and values, and to present a sense of individuality that could help differentiate the organisation within its competitive environment.

[1]Brand identity and brand image are two concepts which are related to each other, yet distinguishable from one another. Relatively, brand image is apparant, passive, backward looking and tactical; brand identity is substantial, active, forward looking and strategic. The former reflects superficial qualities, and focuses more on the receiver's side; the latter reflects enduring qualities, and focuses more on the sender's side.

2.3 *About product identity*

Product identity is a part of brand identity and corporate identity. As the development of brand identity theory, it appeared an emerging awareness of product identity, including such similar viewpoints as product image, product style, product family, product language, brand DNA, etc. Karjalainen (2001) regarded product identity as an insightful and interesting approach to create the trace of brand identity by using a product language for individual products, and proposed a useful and general method to help analyse and build product identity or brand identity.

2.4 *The comparision*

In the initial stage of corporate identity theory, product identity was always present with a part of visual identity, even some practitioners considered it just as the package of product (Selame and Selame, 1988). At that time, brand identity was in the same light as corporate identity. With the market changing constantly, brand identity and product identity are separated from corporate identity gradually.

Usually, it can be divided roughly that product identity does with each product, and brand identity does with a series of products, while corporate identity does with the whole corporation.

3 UNDERSTANDING PRODUCT IDENTITY FROM THREE LEVELS

3.1 *Product as an individual*

A product is the most direct medium for a user to recognize a brand. Through such approaches as form, color, material and interface, the designers send related information like function, quality, style and emotion to the user. As a carrier of experience, it is not only a commodity, but also a bridge between the corporation and the user. In this level, the characteristic of product identity may not be very obvious in every product, but there should be some clues of product identity in its individuality and distinction.

3.2 *Products as a series/cluster*

In the manufacture activities, all of these personalized offerings belong to some product series, or called product lines . Product series is a group of related products having the same prototype and are produced by a single company. Products in the same product series play similar roles, aim at the same target users, or possess a similar market position. It indicates a comprehensive understanding to the target users of the corporation in the product series, which is related to the corporation strategy.

3.3 *Product as a brand*

A product is one of the important carriers of a brand. Users' recognition of a brand may begin with advertisement, communication and activity. And then, these brand impressions will be tested and verified in its products by users.

4 DESIGN IDENTITY SYSTEM MODELS

4.1 *A design identity system integrated with design approach and design thinking*

Among corporate identity, brand identity and product identity, their distinction is the target: the first is to identify different corporations, the second is to identify different brands, and the third is to identify different products. That is, all of them focus on identifying "what". However, design identity focuses on "how" to identify, which takes all kinds of design approaches to serve brand identity, such as product design, communication design, and service design, strategy deign, etc.

According to the theory of the sciences of the artificial (Herbert A. Simon, 1981), brand identity can be regarded as an adaptive mechanism, which can be divided into two levels. One is the exterior level related to the offerings, including product identity, communication identity and service identity; the other is the interior level related to an organization, including mind identity, strategy identity and behaviour identity. As the interface of this adaptive mechanism, design identity is an ensemble of design activities to fulfil brand identity, including product design to product identity, communication design to communication identity, service design to service identity, and mind design to mind identity, strategy design to strategy identity, behaviour design to behaviour identity.

4.2 *A cake model of design identity based on product identity*

Although the design identity system shown is a comprehensive and ideal model, it is too difficult to make clear relationships and direct approaches among these factors. Besides, researches on service design still not goes very deep or sufficient; researches on strategy design and behaviour design just take the first paces by design thinking. (According to three levels of product identity above, integrated with semiology methodology, the significations are divided into denotation, connotation and ideology[2].)

In brand expression and recognition, the key in this model is product identity instead of communication identity. For example, the double-kidney air-inlet grille of BMW is a successful product identity. Besides, brand identity can be built by both product identity and communication identity, which helps to express mind identity of a brand.

---

[2]Based on the general linguistic theory of Ferdinand de Saussure on signifier and signified, Roland Barthes distinguished the denotation and connotation of signification. Furthermore, Fiske and Hartley developed ideology as the third order of signification, which came from Barthes' concept of myth.

In brand management, design identity helps to promote product development from an individual and isolated level to a system level like product series or product cluster, and adjust the relationship among those products in different types and levels of the same brand, so as to control the brand strategy sequentially and gradually.

5 CONCLUSION AND FUTURE STUDY

5.1 *The comparison of design identity system between Sony and Samsung*

In the case study of Sony, there is a high consistency in mind identity, which is both between product series and its brand, and between such sub-brand as Walkman, Cyber-shot and Sony-Ericsson and its mother brand of Sony. In brand management, some product identity features of one brand can be transferred into another related brand, which helps to increase design efficiency and reduce design cost, and simultaneously enhance the relevance of sub-brands so as to strengthen the consistency between sub-brand and mother brand.

In the case study of Samsung, there are not yet some common prototypes in mobile phones, digital cameras and displays. Parts of products such as NV series of blues digital cameras have a consistent image by some curves or elements; however, there is a clear identity context in these products. Product style of Samsung depends more on color collocation and material matching. So, communication identity plays a more important role than product identity in the brand identity of Samsung, which partly attributes to its mass product strategy.

Both Sony and Samsung have planned various product series aiming at different users; and then, each product series follows a specific design code or standard, which is the basis of product identity. Comparatively speaking, there is a more distinct path for Sony to construct brand identity system, which is from mother brand to sub-brands, to product series, and to product individual. From both sub-brands and product series, they support each other in mind identity and product identity, which benefits to take full advantages of corporations. There is not a clear and intentional path for Samsung to construct and control product identity, which probably attributes to its strategy of huge amount of products.

5.2 *Rethinking of design identity cake model*

By the case study of Sony and Samsung, it is proved that the a cake model of design identity is feasible and available. Facing manufacture and aiming at brand identity, this model emphasizes on product identity, which is a bridge from artistic communication identity to abstract mind identity. If communication identity is visible and mind identity is thinkable, product identity is tangible.

To the users, the process of cognizing this model is from communication identity and product identity to mind identity; to the designers and design managers, the process of constructing this model is from mind identity to product identity and communication identity.

5.3 *Future study*

Further study will go along with two ways. One is to extend it from communication identity and product identity to service identity. The other is to extend design identity system along the way of design thinking, which looks for work-arounds and improvised solutions from mind identity to strategy identity and behaviour identity.

REFERENCES

Abratt, R. (1989). A new approach to the corporate image management process. *Journal of Marketing Management*, 5, 63–76.

Albert, S. & Whetten, D. (2003). Organizational identity. In Balmer, J. M. T., Greyser, S. A. Eds.), *Revealing the corporation: perspectives on identity, image, reputation, corporate branding and corporate-level marketing*. Routledge, New York.

Balmer, J. (1998). Corporate identity and the advent of corporate marketing, *Journal of Marketing Management*, 14: 963–996.

Balmer, J. & Soenen, G. (1998). A new approach to corporate identity management, *International Centre for Corporate Identity Studies*, Working Paper, 1998/5.

Blomkvist, J. et al. (2010). This is service design research. in M. Stickdorn, & J. Schneider (Eds.), *This is service design thinking*. Amsterdam, Netherlands: BIS Publishers.

Industrial Engineering and Management Science – Lee (Ed)
 ISBN 978-1-138-02647-6

Predicting the return of the Japanese stock market with artificial neural network

M. Qiu
School of Intelligent Information System Engineering, Fukuoka Institute of Technology, Fukuoka, Japan

Y. Song & F. Akagi
Department of Systems Management, Fukuoka Institute of Technology, Fukuoka, Japan

ABSTRACT: The Japanese stock market is one of the largest stock markets in the world, and the prediction of it is a hotspot research area. As the most widely quoted average of Japanese equities, Nikkei 225 index is a benchmark to value the Japanese economy. Forecasting the stock market returns is a major activity of financial firms and private investors when they make investment decisions. Accurate prediction of the stock market returns is a highly challenging task due to the highly nonlinear nature of the financial time series. In this study, we applied the artificial neural network which can map any nonlinear function without a prior assumption for predicting the return of Nikkei 225 index. Due to the complexity of stock market data, first we selected the 18 input variables from the data of 71 variables that covered financial and economic information of Japanese stock market by the fuzzy surfaces. And then, in order to verify the prediction ability of the selected input variables, we predicted the return of Nikkei 225 index by the artificial neural networks with the learning algorithm of back propagation. For the neural network model, we compare linear regression model with it in the prediction ability of the stock market return. It was observed through empirical experiment that the artificial neural networks performed well, and had a more effective ability than the conventional linear regression in forecasting the Japanese stock market. In addition, the prediction effect of the combination of 18 input variables is effective and can be therefore a good alternative for stock market returns prediction.

Keywords: Finance; Nikkei 225 index; artificial neural network; back propagation

1 INTRODUCTION

In recent years, in order to revive the Japanese economy, the Japanese government makes a lot of significant economic strategies and each strategy is closely related with the Japanese stock market. As the most widely used market index for the Tokyo Stock Exchange, the Nikkei 225 index (also known as Nikkei average or the Nikkei) is a benchmark to value the Japanese economy. Forecasting the stock return of Nikkei 225 index is an important financial subject that had attracted great popularity in major financial markets around the world. It has been widely accepted by many studies that nonlinear exists in the financial markets and that artificial neural network (ANN) can be effectively used to uncover this relationship (Enke and Thawornwong, 2005).

McCulloch and Pitts (1943) created a computational model for neural networks based on mathematics and algorithms. From then on, the study of applying ANN to financial and investment decision has been examined by researchers for many years. Motiwalla and Wahab (2000) found that ANN models were more successful in predicting returns compared to either regression or a passive buy-and-hold strategy. Enke and Thawornwong (2005) used the neural network models for level estimation and classification. They showed that the trading strategies guided by the neural network classification model generate higher profits than any other models. Hodnett and Hsieh (2012) used two learning rules of ANN to forecast the cross-section of global equity returns, their findings support the use of ANN for financial forecasting. In recent studies, ANN is the most popular machine learning methodology and has been proved that the approach can outperform conventional methods (Fernandez Rodrıguez et al., 2000, Versace et al., 2004, Lu, 2010). We applied ANN for forecasting the return of Nikkei 225 index in this study.

In most of the applications, input variables that had been proved to be effective by the previous studies were used to predict the stock market returns. Lagged returns, interest rate value, foreign exchange rate, consumer price index, industrial production index and deposit rate had been effectively used as indicators of stock price (Leung et al., 2000, Enke and

Thawornwong, 2005). In this study, not only the indicators that were proved of the effectiveness by prior studies, we also tried to find some input variables that had not been used to predict the stock market returns before, and check the ability of predicting the stock market index. Therefore, 71 input variables that covered financial and economic information of Japanese stock market were collected, and most of them had not appeared in the previous studies. This study tried to implement the fuzzy surfaces to select the optimal input variables. Finally, 18 good explanatory variables were selected from the 71 input variables to continue the following experiments and most of them had not emerged in previous studies.

Back propagation algorithm is a classical learning algorithm for the neural network, and had been widely applied by many prior studies (Wong et al., 1995). In this study we apply ANN based on BP training algorithm in forecasting the Japanese stock market return. In order to verify the prediction ability of the ANN model, we compare the linear regression model with it in the prediction accuracy of the return of Nikkei 225 index. Experimental results show that the ANN based on BP method improves the prediction accuracy of the return of the Nikkei 225 index and outperforms linear regression model. In addition, the prediction effect of the combination of 18 input variables is effective and can therefore be a good alternative for predicting the stock market returns.

The remainder of this paper is organized into five sections. Section 2 provides the materials and architecture of artificial neural network. Section 3 describes the prediction models. Then we show the experimental results in Section 4. Finally, Section 5 contains the discussion and conclusion. In addition, the data descriptions are given in the Appendix.

2 MATERIALS AND ARCHITECTURE OF ARTIFICIAL NEURAL NETWORK (ANN)

2.1 *Data description*

Nikkei 225 is the most widely used market index for the Tokyo Stock Exchange. In this study, we collected 71 variables that may affect the return of Nikkei 225 index. The entire data set covers the period from November 1993 to July 2013, a total of 237 months of observations. The data set is divided into two periods: the first period runs from November 1993 to December 2007 (170 months of observations) while the second period is from January 2008 to July 2013 (67 months of observations). The first period is used to select the parameters for the forecasting techniques and specifications of the models. The second period is reserved for testing the performance of prediction models using neural networks.

2.2 *The model of ANN*

Funahashi (1989) and Hornik et al. (1989) showed that a neural network with sufficient complexity could approximate any unknown function to any degree of

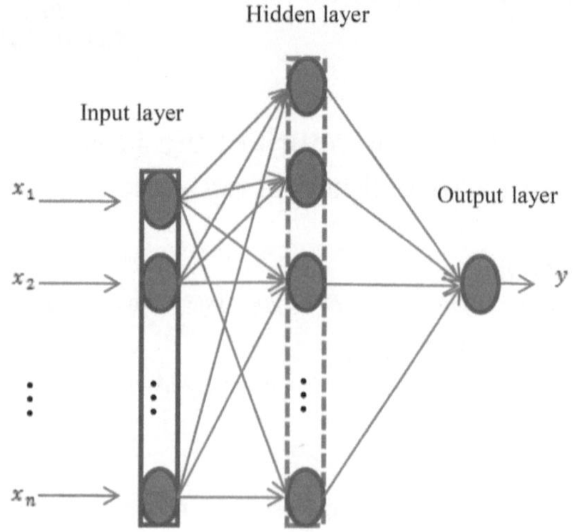

Figure 1. The architecture of three layered feed forward ANN.

desired accuracy with only one hidden layer. Therefore, ANN model in this study consists of an input layer, a hidden layer and an output layer, and each of which is connected to the other. The architecture of ANN is showed in Figure 1. The input layer corresponds to the input variables with one node for each input variable. The hidden layer is useful in capturing nonlinear relationships among variables. We need to find appropriate number of neurons in the hidden layer by repeated training. And the final layer consists of only one neuron that represents the predicted value of output variable.

2.3 *Variable selection*

2.3.1 *Fuzzy surfaces*

Regarding the variable selection, most of the scholars had not carried out the research to identify significant input variables. Some researchers chose the input variables with no explanation, someone directly selected good explanatory variables from the previous studies which had concluded that some variables were effective by using the least squares method, stepwise regression method and even neural network. Here we use a method of fuzzy surfaces which is easily understood and simple to automatically and quickly search the independent input variables for the nonlinear models.

As there are many factors affecting the stock market return, the data in this study has a high degree of non-linear characteristics. We chose fuzzy curve analysis to select the effective input variables of ANN. This method is based on the theory of fuzzy mathematics, and it does not require complicated mathematical modeling. First we calculated the correlation between each input variable and output variable of ANN, and then sorted all input variables according to the importance. There exists relatively large correlation between each input variables, so we excluded the relevant variables

Table 1. Iteration for identifying the important inputs.

	Identified input	Eliminated inputs
Iteration 1	v_6	$v_{37}, v_{42}, v_{41}, v_{35}, v_{48}, v_{47}, v_{65}$
Iteration 2	v_{36}	$v_{38}, v_{51}, v_{19}, v_{43}, v_{34}, v_{32}$
Iteration 3	v_8	$v_{18}, v_{21}, v_{23}, v_{20}, v_{22}, v_{24}$
...	...	...
Iteration 17	v_{55}	v_{28}
Iteration 18	v_{56}	v_{11}

by fuzzy surfaces, and then established the simple and optimal subset of the input variables. The details of the algorithm is shown in Lin et al. (1996).

2.3.2 *The simulation*

Here we use the first period runs from November 1993 to December 2007 (170 months of observations) to select the optimal input variables. The data includes 71 input variables and one output variable. The result is shown in Table 1.

According to the simulation, the significant inputs identified and ranked in order of importance are $v_6, v_{36}, v_8, v_{50}, v_{52}, v_{49}, v_7, v_{14}, v_{12}, v_{17}, v_{44}, v_{30}, v_{10}, v_{33}, v_{54}, v_{53}, v_{55}$ and v_{56}. Input variables which were selected by fuzzy surface are renamed and the meaning of each variable is shown in the appendix. The values of the input variables were first pre-processed by normalizing them within a range of 0 and 1 to minimize the effect of magnitude among the inputs and thus increase the effectiveness of the learning algorithm.

The output variable is the return of Nikkei 225 index, which is computed as follows:

$$y_t = \frac{(p_t - p_{t-1})}{p_{t-1}}$$

where p_t is the value of the index in month t, p_{t-1} is the value in month $t-1$, dividends are ignored for this study.

According to the selected input variables, we found that some variables, such as T-bill rate, had been proved of the effectiveness and often been used by the previous studies. However, most of them had not been examined before, in this study we will verify the predicted effects of these variables in the following models. In addition, due to the lag associated with the publication of macroeconomic indicators, we set certain data with a one-month time lag. And it will be similar to real-world practice by using these variables in the forecasting models.

3 PREDICTION MODELS

3.1 *Back propagation neural network training*

3.1.1 *Back Propagation (BP) in ANN training*

After the introduction by Werbos (1994), Back propagation (BP) was popularized as a supervised learning algorithm for the neural network. BP algorithm is a classical learning algorithm for the neural network, and had been widely applied by many prior studies. The main idea of BP algorithm is; enter the study samples, and then we can use the BP algorithm to adjust the weights and bias of network by repeated training. In such a way that the error between the desired output and the actual output is reduced. When the error is less than a specified value or the termination criteria is reached, we can complete the training and save the weights and bias of the network.

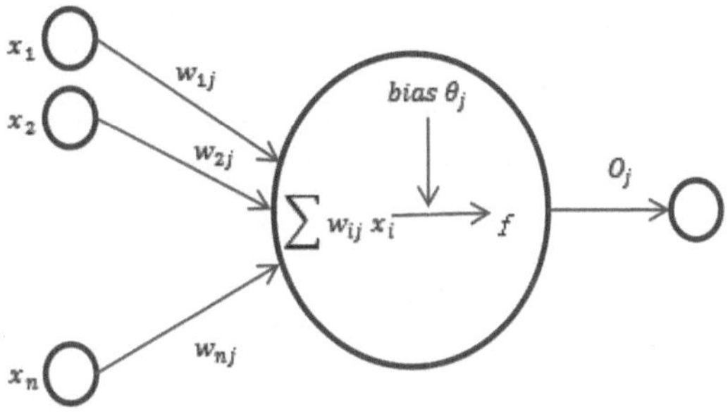

Figure 2. Processing (hidden or output) unit j.

Simple algorithm steps are shown as follows, and Figure 2 displays the illustration of an artificial neural network processing unit (Han and Kamber, 2006, Kantardzic, 2011).

Step 1 Initialize the connection weights (w_{ij}) between ith neurons in the previous layer and unit j and the biases (θ_j) at random for each processing (hidden or output) unit j.

Step 2 Propagate the inputs forward

(1) Compute the input of unit j, (I_j) with respect to the previous layer as follows:

$$I_j = \sum_{i=1}^{n} w_{ij} x_i + \theta_j,$$

In the formular, x_i is the output of ith neurons in the previous layer.

(2) Calculate the output (O_j) of each unit j by using an activation function (in this study, we use sigmoid function).

$$O_j = \frac{1}{1+e^{-I_j}}.$$

Step 3 Back propagate the errors

For each unit j, the adjustable error is calculated as follows:

(1) If node j is an output node, then we note

$$Err_j = O_j(1-O_j)(T_j - O_j),$$

where O_j is the actual output of unit j, and T_j is the true output based on the training sample.

(2) If node j is an iternal hidden node, then we define

$$Err_j = O_j(1-O_j)\sum_{k} Err_k w_{jk},$$

Table 2. ANN parameter levels tested in the experiments.

Parameters	meaning	Level(s)
n	The number of neurons in the hidden layer	10, 20, ..., 100
ep	number of iterations	1000, 2000, ..., 10000
mc	momentum constant	0.1, 0.2, ..., 0.9
lr	value of learning rate	0.1

where w_{jk} is the connection weight from unit j to a unit k in the next higher layer, and Err_k is the error of unit k.

Step 4 Updating of the weights and biases.

(1) For each weight w_{ij} in the network, we calculate

$$\Delta w_{ij} = (lr)Err_j O_i,$$

$$w_{ij} = w_{ij} + \Delta w_{ij},$$

where Δw_{ij} is the change in weight w_{ij}. The variable lr is the learning rate, a constant typically having a value between 0.0 and 1.0.

(2) For each bias θ_j in the network, we note

$$\Delta\theta_j = (lr)Err_j,$$

$$\theta_j = \theta_j + \Delta\theta_j,$$

where $\Delta\theta_j$ is the change in weight θ_j.

Step 5 Select the next pair of input patterns and then train the network repeatedly according to Step 2. Training stops when: (1) all Δw_{ij} in the previous epoch were so small to below some specified threshold, or (2) the percentage of samples misclassified in the previous epoch is below some threshold, or (3) a prespecified number of epochs has expired.

3.1.2 *The experiments*

Since there are several parameters that influence the predictive ability of ANN model, we experiment a parameter setting of ANN based on BP algorithm to find the most appropriate one. In addition, we can also find some characteristics of BP algorithm according to a large number of experiments. Ten levels of n, nine levels of mc and ten levels of ep were tested in the following experiments. As suggested in the previous literatures, a small value of lr was selected as 0.1. The ANN parameters and their levels are summarized in Table 2.

The experiments were completed by a total of 900 parameter combinations. And we select the parameter combination that give the best performance. In this study, we use the mean square error (MSE) to evaluate the performance of the ANN model, the MSE is defined as follows:

$$MSE = \frac{1}{N}\sum_{t=1}^{N}(y_t - \hat{y}_t)^2,$$

where y_t denotes the actual return of Nikkei 225 index, and $\hat{y}_t$ is the predicted return we calculated.

After we finished a total of 900 parameter combinations, we found that the most appropriate parameter combination in this study is ($n = 10$, $ep = 3000$, $mc = 0.4$, $lr = 0.1$). And the value of MSE for the best BPNN model is 0.0017. The average value of MSE of the 900 times training experiments is 0.1077.

During the experiment, we found several characteristics:

- With the increase of the number of neurons in the hidden layer, the speed of the computer operation become slower, but the value of MSE does not decrease gradually.
- The parameter combination with the relatively smaller value of MSE, always has the relatively less number of neurons in the hidden layer. The value of MSE is relatively smaller when the number of neurons in the hidden layer is with the value between 10 to 30 in the experiments.
- We may got different values of MSE for each iteration when ANN was trained by the same parameter combination. So we just do one time for every parameter combination.
- When the number of the neurons in the hidden layer become larger, the computer may cost longer time to capturing nonlinear relationships among variables. But for many parameter combinations with large number of neurons in the hidden layer, the experiment was ended in a short time. We speculated that the experiments achieved the best solution in the region of their starting point which is the local minimum.
- Due to the complex and huge data, the time for converging was so long that we may need about 1 hour for one parameter combination.

3.2 *Linear regression model*

In order to verify the prediction ability of neural network model, we use the method of backwards for establishing the linear regression model to compare with it. This method started with the full set of the selected 18 input variables, and then removed the variable which had the least contribution to the model from the last updated set. The final model was generated until the deletion of any variables would significantly affect to the model. The significant t-test was used as criteria of the significant input variables in the linear regression model. The remaining variables were thus used to predict the stock market returns. In this study we kept $x_2, x_3, x_8, x_9, x_{13}$ and x_{15} as the significant input variables in the regression model. The regression model has the following function:

$$R_t = 0.000000764x_2 - 0.00003182x_3 + 0.005506x_8 + 0.004942x_9 - 0.002047x_{13} - 0.03664x_{15} + 1.095 + \varepsilon$$

Here ε is the error term that follows the normal distribution. In this regression model, all the regression

Table 3. Error analyses of different forecasting models.

Models	LR	BPNN best	BPNN average level
MSE	12.7800	0.0044	0.1219

coefficients are significant and the F-statistic is 3.388 (p-value $0.004 < 0.05$), indicating that these forecasting variables can reflect the information of the stock market returns. The regression model shows that the changes of x_2, x_8 and x_9 have a positive effect on predictions of stock returns, whereas the effect on stock market return of x_3, x_{13} and x_{15} is negative.

4 EXPERIMENTAL RESULTS

In this study, The models were tested on Windows 7 operating system, and we applied MATLAB R2011a by Math Works for operating all the experiments. Each of the models described previously is estimated and validated by the in-sample data. At this stage, the empirical evaluation for each model is based on the untouched out of sample data which covers from January 2008 to July 2013 (67 months of observations). This is due to the fact that the superior in-sample performance does not always guarantee the validity of the forecasting accuracy.

With the formulas of *MSE* above, the performance evaluation and the comparison of different models are calculated in Table 3.

LR denotes the model of linear regression. Since 900 times of experiments on BP traning had been executed in Section 3.2, here we chose the best ($n = 10$, $ep = 3000$, $mc = 0.4$, $lr = 0.1$) parameter combination of BPNN (Back propagation neural network) model to compare with LR model. The average level of the performance of BP algorithm is also shown in Table 3.

In the Table 3, the smaller the criteria are shown, the better the prediction effect. From Table 3, we can see that LR model has the largest value of *MSE*. The value of *MSE* for the average level of BPNN model is 0.1219. The best model which we found from the large number of experiments performs well and the value of *MSE* is 0.0044. We conclude that the prediction ability of the BPNN model is much more effective than the performance of the linear regression model. From the performance of the experiments, we also found that even though the 18 effective input variables had not been examined, the prediction effect is still effective. The combination of 18 effective input variables can be therefore a good alternative for stock market returns prediction.

5 CONCLUSIONS

In this study, in order to search more new effective input variables, which are used to predict the return of Nikkei 225 index, we collected 71 variables that refer to different aspects of the Japanese stock market. And then we selected new combination of input variables of 18 good explanatory variables by fuzzy surfaces and utilized the combination to predict the return. We used the monthly data of 18 good explanatory variables to predict the return of Nikkei 225 index, and then compared the ability of the prediction for different models. We found that the forecasting accuracy of ANN based on BP algorithm is much more effective than the conventional linear regression. In addition, the prediction effect of the combination of 18 input variables is effective, and we may apply it for other stock markets for the future research.

Appendix: Meaning of the renamed input variables.

Input variables (renamed)	Meaning
x_1	Average Amounts Outstanding of Monetary Base
x_2	Banknotes in circulation of average amounts outstanding of monetary base
x_3	Coins in circulation of average amounts outstanding of monetary base
x_4	Uncollateralized overnight of call rates at the end of month
x_5	Yen spot rate at the end of month of Tokyo market
x_6	Yen central rate at the end of month of Tokyo market
x_7	Yen lowest in the month of Tokyo market
x_8	Percent changes from the previous year in average amounts outstanding of money stock
x_9	Percentage changes in average amounts outstanding from the previous year of loans and discounts for total of major and regional banks
x_{10}	Loans and discounts of regional banks
x_{11}	Import price index of all commodities
x_{12}	Real exports
x_{13}	Real imports
x_{14}	Indices of industrial production
x_{15}	1-year T-bill rate
x_{16}	2-year T-bill rate
x_{17}	3-year T-bill rate
x_{18}	4-year T-bill rate

REFERENCES

Enke, D. & Thawornwong, S. 2005. The use of data mining and neural networks for forecasting stock market returns. *Expert Systems with Applications*, 29, 927–940.

Fernandez Rodrıguez, F., Gonzalez Martel, C. & Sosvilla Rivero, S. 2000. On the profitability of technical trading rules based on artificial neural networks: Evidence from the Madrid stock market. *Economics Letters*, 69, 89–94.

Funahashi, K. I. 1989. On the approximate realization of continuous mappings by neural networks. *Neural Networks*, 2, 183–192.

Han, J. & Kamber, M. 2006. *Data Mining, Southeast Asia Edition: Concepts and Techniques*, Morgan kaufmann.

Hodnett, K. & Hsieh, H. H. 2012. Application of cascade-correlation neural networks in developing stock selection models for global equities. *International Business & Economics Research Journal* 11, 375–396.

Hornik, K., Stinchcombe, M. & White, H. 1989. Multilayer feedforward networks are universal approximators. *Neural Networks*, 2, 359–366.

Kantardzic, M. 2011. *Data mining: concepts, models, methods, and algorithms*, John Wiley & Sons.

Leung, M. T., Daouk, H. & Chen, A.-S. 2000. Forecasting stock indices: a comparison of classification and level estimation models. *International Journal of Forecasting*, 16, 173–190.

Lin, Y., Cunningham III, G. A. & Coggeshall, S. V. 1996. Input variable identification—fuzzy curves and fuzzy surfaces. *Fuzzy Sets and Systems*, 82, 65–71.

Lu, C. J. 2010. Integrating independent component analysis-based denoising scheme with neural network for stock price prediction. *Expert Systems with Applications*, 37, 7056–7064.

McCulloch, W. S. & Pitts, W. 1943. A logical calculus of the ideas immanent in nervous activity. *The Bulletin of Mathematical Biophysics*, 5, 115–133.

Motiwalla, L. & Wahab, M. 2000. Predictable variation and profitable trading of US equities: a trading simulation using neural networks. *Computers & Operations Research*, 27, 1111–1129.

Versace, M., Bhatt, R., Hinds, O. & Shiffer, M. 2004. Predicting the exchange traded fund DIA with a combination of genetic algorithms and neural networks. *Expert Systems with Applications*, 27, 417–425.

Werbos, P. J. 1994. *The roots of backpropagation: from ordered derivatives to neural networks and political forecasting*, Wiley.com.

Wong, B. K., Bodnovich, T. A. & Selvi, Y. 1995. A bibliography of neural network business applications research: 1988–September 1994. *Expert Systems*, 12, 253–261.

Industrial Engineering and Management Science – Lee (Ed)
© 2015 Taylor & Francis Group, London, ISBN 978-1-138-02647-6

An information system of community home-based elderly service and its application on curriculum design

JunHua Zhang
Ningbo City College of Vocational Technology, Ningbo, China

ABSTRACT: With the rapid increase of elderly population, caring for the elderly is becoming a big and hot social issue. Community Home-based Elderly Service (CHES) is one of the mainstream ways to solve the problem. In order to support the CHES, an information system is introduced in this paper. The functions, database design, framework design and concrete implementation of the system are presented. Furthermore, following the direction of curriculum reform for high vocational education, using real projects as the basic content of curriculum is encouraged. In this paper, the above system is parsed according to its professional coding skills, and the route of design a new course is depicted.

1 INTRODUCTION

With the rapid increase of elderly population, how to care for the elderly is a big and hot social issue. Traditional nursing the aged is depended on family, which is difficult to continue as family structure changes. Community Home-based Elderly Service (CHES) is raised as a new and effective way. Information technology can provide real-time, cheap and colorful service for the aged and their supervisor. This paper will introduce a succinct information system developed for the goals by our team.

Paper [1] sums up the service modes for CHES, and suggests the concept of virtual nursing home and its service objects, contents and ways. Paper [2] presents a design scheme of an information system for home nursing of the aged, its main subsystems and their implementation.

The system developed by our team is similar to [1] and [2], the aim is also to provide a effective information system for CHES. But our system is better satisfied with concrete needs for real CHES. For example, the systems in paper [1] and [2] are based on many special hardware equipments, such call center, database server, disk matrix, and professional client software, etc. In real CHES, there are not so much calls in a same time, and also these equipments can be instead by telephone, QQ and wechat. Our system runs on common personal computers and internet, and is suitable for the information management of CHES.

Paper [2] is constructed by a heavy framework J2EE, but our system is developed just using the combinations of jsp, servlet and javabean. The reason is, the information processing tasks are not busy enough to use the heavy framework, which make the system difficult to develop, maintain and run fast.

With the progress of curriculum reform in high vocational education, people realize that it is a good practice to use a real and integrated project as the backbone to construct a curriculum of program design. In this paper, we will parse our information system for CHES and give a route of how to design a course of "JSP development and application" with also modern vocational education thoughts.

2 AN INFORMATION SYSTEM OF COMMUNITY HOME-BASED ELDERLY SERVICE

2.1 *System functions*

Through a lot of investigation and study, we determine the requirement for the information system of CHES. Fig. 1 presents the functions of the system.

Among the functions, "Work Order Create" is used to create a new Work Order. A Work Order contains No., social worker's name, start time of service, end time of service, guest's name, service item, service fee, whether fee is paid or not, appraise to service. "Work Order Close" is used to fill up the items blank at generation, such as end time of service, appraise to service. "Work Order Query" is used to inquire about Work Order. "Guest Fee List" is used to generate guest fee sheets according to different demands. "Other Query" is used to inquiry all of social workers, guests, and monthly total fee.

The system is served to ordinary guests and managers of CHES, especially for the latter. When ordinary guests access the system, they can see several kinds of messages. When managers access the system, they need login the system, and operate the system with the functions provided as Fig. 1.

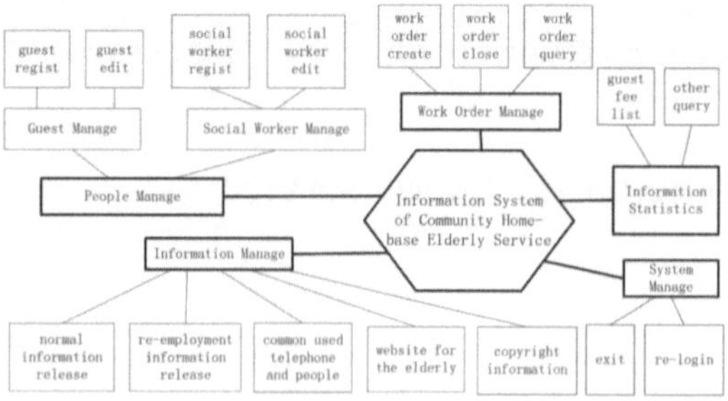

Figure 1. Functions of the information system for CHES.

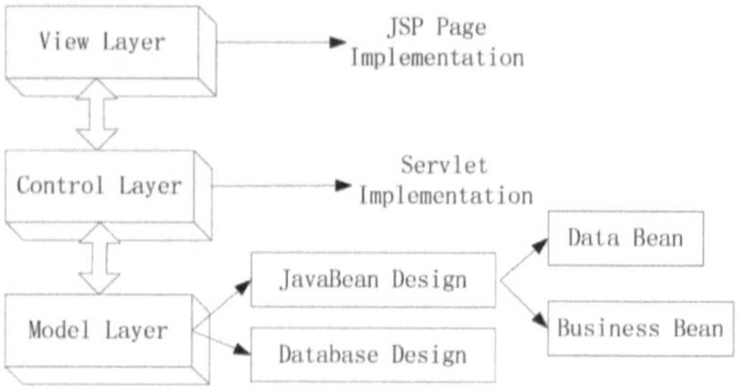

Figure 2. System whole framework.

2.2 *System framework*

The system of CHES adopts a framework of three layers, which are model layer, view layer and control layer. Model layer pays only attention to data storage. View layer focus on display of content. Control layer coordinates model layer and view layer, and implements business process.

According to user and requirement of information process, we adopt JSP technology as the program language and MyEclipse as the development platform. Through combination of jsp, servlet and javabean in JSP technology, the MVC framework can be implemented. Fig. 2 is our system framework [3].

In order to provide JSP page with good quality, the web page tool DreamWeaver is used. The results from DreamWeaver can be input into MyEclipse as the basis for continual coding.

2.3 *Database design*

According to the system's functions, data source is analyzed and database is planned. Through data analysis, reasonable table structure and table relationship is designed. Concrete works contain database requirement analysis and database logic structure design.

MySQL is adopted as our system's database. In our database, there is social worker table, old man table, Work order table, message table. For example, there are items of social worker No., social worker name, sex, ID card no., native place, address, telephone, work sort, salary, start work time and level in the social worker table.

2.4 *System design in detail and implement*

For every system function, according to the system framework, jsp, servlet and javabean files from model layer, view layer and control layer are designed, coded and debugged. Increment test continues during the development process. At last, whole test is done, in which errors and defects are found and removed.

3 THE APPLICATION OF THE INFORMATION SYSTEM FOR CHES IN CURRICULUM DESIGN

3.1 *Teaching thought*

Constructivism teaching design rule proposes that student's study must be combined with some tasks or problems; Learner's study interesting and enthusiasm is inspired and maintained by probing questions; Create real study tasks; Students must occupy study's initiative, and teachers must keep encouraging student's progress [4]. Study habits, basis and intelligence characters of students in high vocational education determine that constructivism is more suitable for these students' teaching. Besides satisfying this rule, there are two other questions to be solved.

The first question is, how to apply the material of an information system to curriculum design, so that students no only learn some knowledge, but also accumulate experience of developing an information system. Of course, if we can let professional knowledge and technology applied relative uniform in the chapters of a curriculum, it will be better.

The second question is, for every professional technology, students will suffer the stages from imitate, understand to grasp. So it is better that the teaching steps in a learning cycle is divided into three phases: Imitate, Analyze and Innovate [5].

In the stage of Imitate, teacher displays the running effect. Students input codes into computer and let the system run. In the stage of Analyze, students are asked to modify the codes, which make them understand the meaning of the codes. In the stage of Innovate, students are required to develop a similar new function. After the training of IAI cycle, students will grasp a new knowledge or technology effectively.

In this paper, the contents of the information system for CHES will be arranged properly, and a new course is designed for students to study the behind knowledge and technology integrated with above teaching thoughts.

3.2 *Parse of the information system for CHES*

We have parsed the information system for CHES, and the result is presented in table 1.

3.3 *Organize and arrange of the information system for curriculum design*

In the curriculum design, the chapters should correspond to the functions in the information systems.

Table 1. System's functions and new knowledge & technology contained.

No.	Function Point FN	New Knowledge (KN) and Technology SL	Software Engineering Jobs
1	Prepare for develop	**SL01**: set up develop environment	Requirement analysis and database design
2	Total Index page and manager login	**SL02**: draw total index page using tool DreamWeaver **SL03**: create an JSP file in MyEclipse, and import the result of SL02 **KN01**: concept and process flow of JSP+SERVLET+JAVABEAN **SL04**: create and information process of Servlet **SL05**: create and information process of Javabean (for database)	Design in detail for manager login
3	Index page of manager	**SL06**: whole design for manager operation using framework set of DreamWeaver	Design in detail menu of index page of manager
4	Guest register	**KN02**: request **SL07**: application of request	Design in detail: generate guest no., pass guest no through request, display register page
5	Guest edit	**SL08**: Operation of database (create connection, query and update)	Design in detail: input guest no., query the detail of a guest and display
6	Social worker register	**SL09**: create and apply of Javabean (for public information process) **SL10**: operation of database (save)	Design in detail: process flow when social worker no. tautology
7	Social worker edit	**KN03**: response **SL11**: application of response	Design in detail: input social worker no., query the detail of the social worker and display
8	Work order create		
9	Work order close	**SL12**: debug skills in MyEclipse	
10	Work order query	**SL13**: process of Chinese words chaos **SL14**: message obtain and process of a jsp page	Design in detail: design a group of query conditions and do the query
11	Guest fee list	**SL15**: application of JavaScript for input legal check **KN04**: session **SL16**: application of session **KN05:** out **SL17**: application of out **SL18**: display of records in different pages **SL19**: Operation of files	Design in detail: design a group of query conditions and do the query
12	Other query		
13	Message release	**SL20**: usage of List frame	Design in detail: design the page of message release
14	Message edit	**SL21**: usage of the variant List	Design in detail: inquiry, display, edit and delete of message
15	System logout		Design in detail: how to set the property of session
16	System re-login		
17	Total index page: message display	**SL22**: design of the total index page especially for message display	Design in detail: design of the total index page and the scheme of message display

While new knowledge and technology need be used, they will be introduced carefully at first, and then be applied in real program design and coding.

According to IAI rules, if a kind of knowledge or technology is used initially, we can give the whole design or code to use it; when it is used later, we just give the information process flow, students are asked to finish the detailed design or code themselves, or detailed design or code are presented but errors contained, students need debug the code and remove the

Table 2. A scheme of curriculum design for the course of "JSP development and application".

No.	Chapter	Functions contained
1	System development prepare	FN1
2	Index page design	FN2, FN3
3	Guest information process	FN4, FN5
4	Social worker information process	FN6, FN7
5	Work Order information process	FN8, FN9, FN10
6	Information statistic	FN11, FN12
7	Information release and edit	FN13, FN14
8	System manage	FN15, FN16, FN17

errors. For example, in table 1, FN8 does not involve new professional knowledge or technology, the code should be done by students themselves. In FN9, we can give a part of code containing several errors.

Combination of the system content parsed in table 1 and the curriculum design thoughts discussed above, we can present a scheme of curriculum design for the course of "JSP development and application" as table 2.

For every chapter, there are four steps to do: step one, design of information process flow; step two, interface design; step three, introduce new knowledge and technology; last step, finish the work in this chapter. The content of step one is provided by teaching material directly, so the coming design and coding could be unified. In step two, students are asked to grasp the basic elements of interfaces; the use of CSS is optional. The content of step three is also provided by teaching material whole, which are high effective material (if no new knowledge or technology is involved, this part can be ignored). In step four, teaching material just provide a part of code as clue, others are done by students themselves as discussion above.

4 CONCLUSION

The information system for Community Home-based Elderly Service is a pure software system based on the technology JSP, and no special hardware is needed. This let the system be applied with low cost. The system can run effectively with clear construction, easy maintenance and extension.

Based on this information system, we depict a new curriculum design of "JSP development and application". The chapters of the course correspond to functions of the system. New knowledge and technology are introduced in their first application, and consolidate in the latter chapters. When the course is studied over, students not only grasp the main knowledge and technology of JSP, but also accumulate the real experience to develop a project based on JSP.

ACKNOWLEDGEMENT

This research is supported by Science Research Foundation of Zhejiang Educational Committee (No. Y201430684) and Foundation for Professional Leader of Zhejiang High Vocational Education (No. lj2013174).

REFERENCES

[1] Zuo Xianlan and Zhang Junhua. Virtual Nursing Home for the Aged: To Update the Community Home Nursing for the Aged [J]. Reformation and Strategy, 2013, 29(9):114–118. (in Chinese)

[2] Ou Yu. Design and Implementation of Home Care Service Information System [D]. Beijing: Beijing University of Posts and Telecommunications, 2010:a16–23, b24–25. (in Chinese)

[3] Kong Hao and Lu Xia. Java Web Design Practical Teaching Material—The Application Development based on Eclipse [M]. Beijing: China Machine Press, 2012: 116–121. (in Chinese)

[4] Wang Xueqing. Project Based Teaching Probe and Practice on Program Design Course [J]. Vocational Education Research, 2009(2):88–89. (in Chinese)

[5] Jiang Xiaogang. Discuss of Practice Teaching on Software Technology Profession of High Vocational Education [J]. Vocational Education Forum, 2010(17): 20–22. (in Chinese)

Industrial Engineering and Management Science – Lee (Ed)
© 2015 Taylor & Francis Group, London, ISBN 978-1-138-02647-6

Multiple-channel and intelligence digital diagnostic system for SFP optical transceiver

Shuai Liu, Hui Hong & FeiHu Song
Institute of Microelectronic CAD Hangzhou Electronic University, Hangzhou, China

ABSTRACT: In this paper, a multiple-channel and intelligence digital diagnostic system for Small Form-factor Plug-gable (SFP) optical transceiver is presented. A few parameters which are regulated by the agreement of SFF-8472, can be monitored and displayed in real time, as well as engender appropriate warnings and alarms. Some measurement engineers and users could input data offline and modify the correlative data at the address of A0h and A2h. A new multi-machine communication structure is also proposed, while the slaves could communicate with a host, whether they have the same device-address or not. As the result, multiple SFP optical transceiver modules could be monitored at the same time.

1 INTRODUCTION

In recent years, optical fiber communication has made rapid development. Optical transceiver module (OTM) as a key component of fiber optic communication network is widely used in a variety of optic-fiber communication systems, and various networks need more and more types of optical fiber modules with function increasingly demanding. To meet the growing demand, OTM is developed toward the high-speed and intelligent direction. Intelligence OTM, which includes the function of Digital Diagnostic Monitoring (DDM), is regarded as an iconic product by many manufacturers who wanted to upgrade their technology in SFP optical transceiver modules.

The function of DDM is described by the agreement of SFF-8472. The temperature, supply voltage, bias current, TX power and RX power of OTM should be monitored, and the DDM allows the designers to find the wrong point in optical fiber link at first time. This way is to simplifying the maintenance service and improving the reliability of the whole network system. Accompanied by the continuous development of the short distance MAN and another networks, the requirement of optical transceiver module will have a significant increase. Intelligence OTM will become the mainly trend. Simultaneously, offline monitoring for OTM at real time is becoming more and more important.

2 AGREEMENT

On the basis of the agreement of SFF-8472, the system divides the address and content into many parts which is shown in Figure 1. The product and manufacturer information is included in the register where

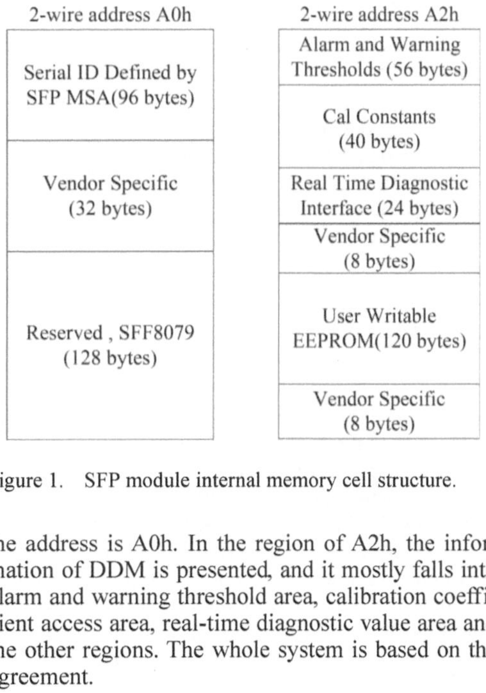

Figure 1. SFP module internal memory cell structure.

the address is A0h. In the region of A2h, the information of DDM is presented, and it mostly falls into alarm and warning threshold area, calibration coefficient access area, real-time diagnostic value area and the other regions. The whole system is based on the agreement.

3 INTEGRAL STRUCTURE

The whole topology of system is shown in Figure 2. The BD_ONE and BD_TWO circuits are Tong-links which provide hardware environment to OTM, and the BD_THREE circuit is an assembly which is composed by Module Select Unit (MSU) and DDM System unit. The SerDes (Serializer/Deserializer) transforms parallel signals which produced by FPGA to the high speed

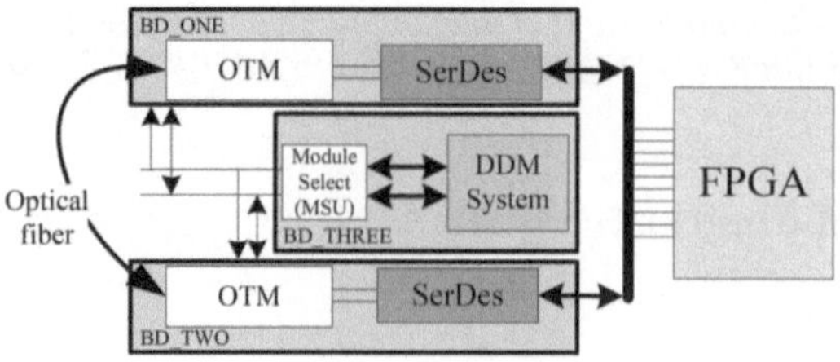

Figure 2. The whole topology of system.

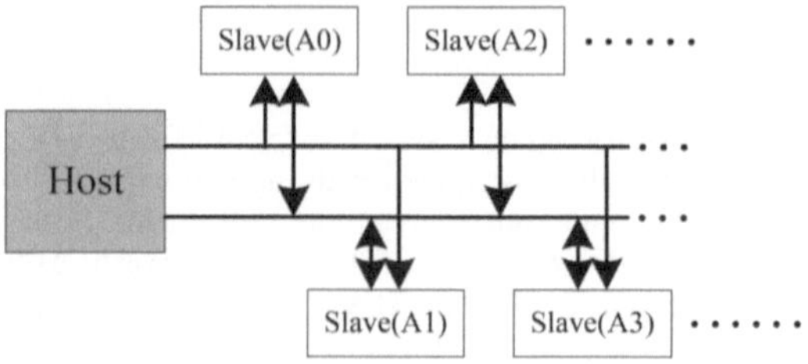

Figure 3. The topology structure of Traditional 2-wire Multi-Machine Communication (TMMC).

differential signals and the OTM module converts electrical signals into optical signal which is spread to the Rx-receiver by an appropriate optical fiber. According to the receiver, it conducts photoelectric conversion. Besides, the connection between the MSU module and the OTM module conforms the 2-wire communication interface which is prescribed by the SFF-8472 agreement, as well as DDM and MSU are connected with GPIO, and then monitor the different OTM.

4 BLOCK CONSTRUCTION

4.1 *MSU*

For general communication bus, as SPI, IIC and so on, no matter the point-to-point which means a host to a slave or multi-machine communication which means a host to a lot of slaves. From their communication protocols, the hosts find the slaves device-address first, and then send much start information, data information and ending message to the slaves. There is a topology structure of traditional 2-wire multi-machine communication (TMMC) in Figure 3. But the biggest disadvantage is that TMMC can not support the slaves which have the same device-address.

The same device-address of the OTM is A0h and the region of DDM is A2h which is shown in Figure 1. Figure 4 describes the communication network, which the SFP1, SFP2 and SFP3 are the OTM. They have same device-address and the circular ring for a host. While the host sends a message of A0h to the 2-wire bus, regrettably, the host does not know which slave has the device-address of A0h, and then the communication mechanisms can not be established between them. So monitoring Multi-OTM at one time cannot be achieved if TMMC has been used.

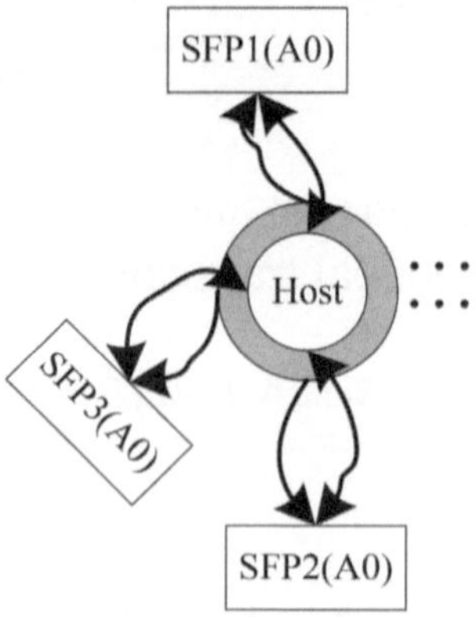

Figure 4. Multi-OTM communication network which has the same device-address.

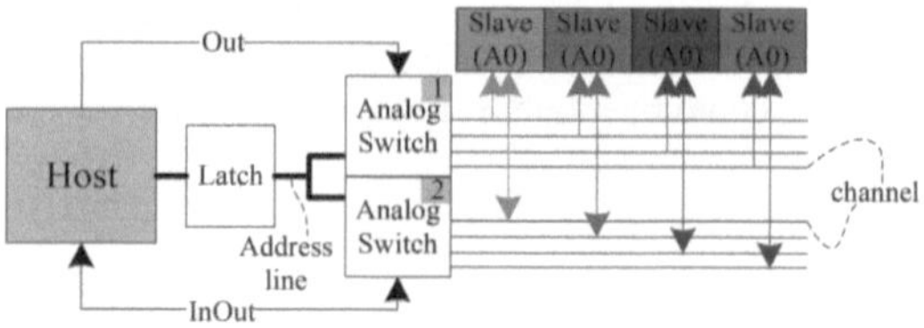

Figure 5. A brand-New Multi-Machine Communication (NMMC) structure.

For this problem, the system put forward a bran-new multi-machine communication (NMMC) structure, as shown in Figure 5. It introduced multi-channel analog switch and latch on the basis of TMMC. Among them, all of slaves have the same device-address (A0h) and they are attached to some different channels which are in/out ports of analog switch, and the channel can be changed by appointed address line of this integrated circuit.

To-warding to the 2-wire, there are two analog switches which must be used, and some reasons will be given. At first, the data and clock is detached, and some pull-up resistors must be connected between the power and signal wire according to their communication protocol. The implementation of the IIC agreement is an example. If an analog switch which controls the clock is thrown away, the slave clock and pull-up resistor will be paralleled together separately. It is shown in Figure 6, and this circuit will cause the total pull-up resistor that is much different from the specified value, so they cannot communicate with each other.

In summary, two analog switches are needed to control the bus separately at this situation. There is a conclusion for NMMC; the best way is to introduce N-switches for an N-wire communication system, if the system does not take pull-up resistor into account. The switch should be removed on the clock line path.

4.2 *DDM unit*

The DDM unit is used to disposal some monitoring parameters that are transferred by the MSU. It is

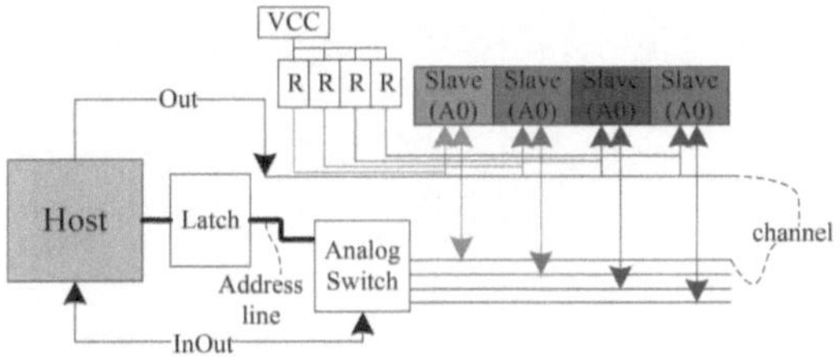

Figure 6. The structure with one analog switch.

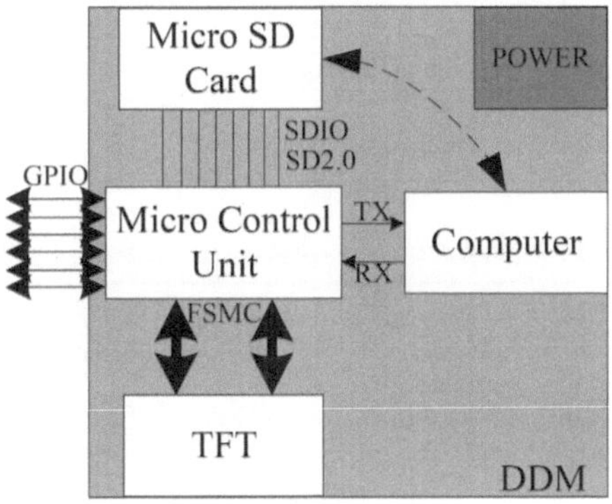

Figure 7. The topology of DDM unit.

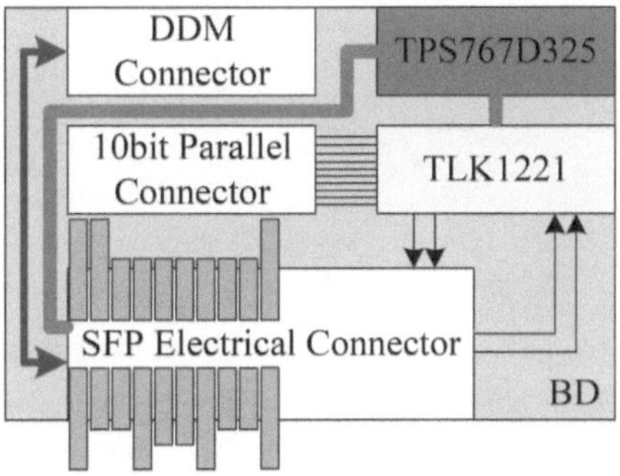

Figure 8. Platform of optical fiber communications.

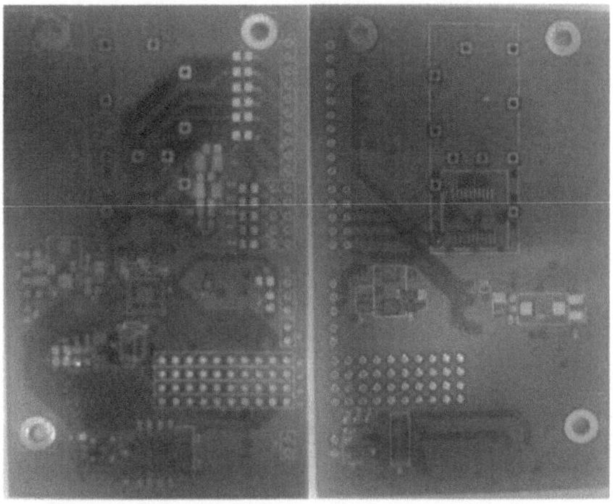

Figure 9. Print circuit board of platform.

composed by some segments which is shown in Figure 7. The MSU and DDM unit are communicated with each other by the Micro Control Unit GPIO which is abide by 2-wire timing sequence of coding, the intelligence and no computer-aided function to input the data is the aim of the system.

The Micro Control chip STM32F103ZET6 which is a 32-bit processor is been used to control the other modules which are included by DDM. Micro SD card is used for logging data offline which is stored in A0h and A2h of OTM, and it also follows the agreement of SD2.0. In order to exchange Big-data which means to deal with a lot of OTM at one time and manage files well, the FATFS that is an embedded file system is transplanted into DDM. As well as the processed values are displayed on touch screen which sustains the interface of RGB6/16/18bit, FSMC is a bridge between TFT and Micro Control Unit rather than SPI to change the screen page very fast and up-data data quickly.

4.3 *Platform of optical fiber communications*

According to SFP_MSA, the OTM circuit has the standard electrical interface, and then a platform for optical fiber communication in Figure 8 is designed. The TPS767D325 chip is a twin channel LDO, which outputs the voltage of 2.5 V and 3.3 V to supply the whole power for this system. The TLK1221 chip is a SerDes, and it supports data rates from 0.6 Gbps through 1.3 Gbps. The 10 bit parallel signals are converted into a high-speed differential signal as the TOSA input that is included in OTM, and thereby it generates the optical signal. The platform also contains many other interfaces to communicate with another unit. Considering the system integrity of power and signal, a 4-layers print circuit board designed by Cadence Allegro 16.5 is implemented (shown in Figure 9).

5 THE SOFTWARE ALGORITHM DESIGN

The TFT display is adopted to realize human-computer interaction in this system, and the interface includes five paginations: the real time page, the warn page, the alarm page, the A0h region page and the channel select page. One page means one full screen. A software flow pattern is shown in Figure 10. As the power on, the whole system starts running, the trigger is valid for real time page, and the collected data from the OTM circuit will be calibrated according to SFF-8472 calibration methods. The OTM internal calibration algorithm is designed. For example, the temperature parameter includes positive and negative values (yielding a total range of −128°C to +128°C), and it is given by the signed two complement values with LSB equal to 1/256°C. The code flow is shown in Figure 11 and then the actual value is presented in TFT. In addition, 256 bytes in the region of A0h will

Figure 10. The whole software flow pattern.

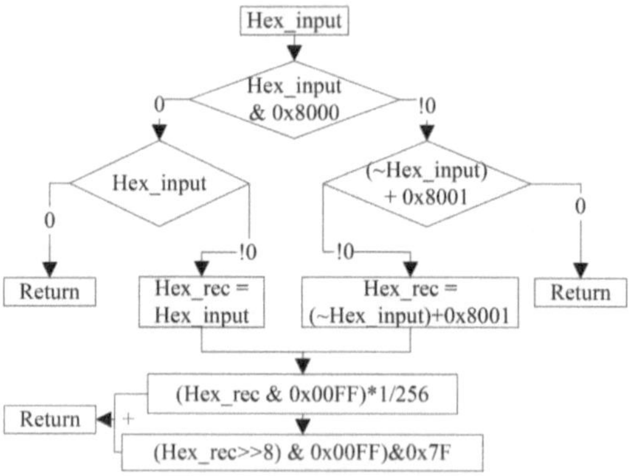

Figure 11. Calibration algorithm for OTM temperature.

be displayed, but they can not be displayed completely in one page. The A0 region page is divided into two parts and they are 0–129 Bytes and 130–255 Bytes respectively.

6 EXPERIMENTAL RESULT

The OTM from different companies to access our system is chosen in experiment. The following Table 1 shows the parameters of OTM. Simultaneity, the double LC interface and the multimode optical fiber (the length of 3-meters) had been used. Also we can see the whole system in Figure 12.

Figure 13 shows the parallel output of the SerDes with one pin. While the FPGA module sends 10bit parallel signals to SerDes input, its direct current keeps balance. For example, the coding is 10'b1010101010,10'b0011101010 and so on. The Figure 13(a) shows the result is the normal condition; when disabled OTM by its tx-disable pin which is defined by SFP_MSA, phenomenon occurred like picture two. The oscilloscope MSO7054B is used in the experiments.

Some monitoring values in TFT as shown in Figure 14 are got and all of the parameters are come from different OTM. In addition, a region for the voltage

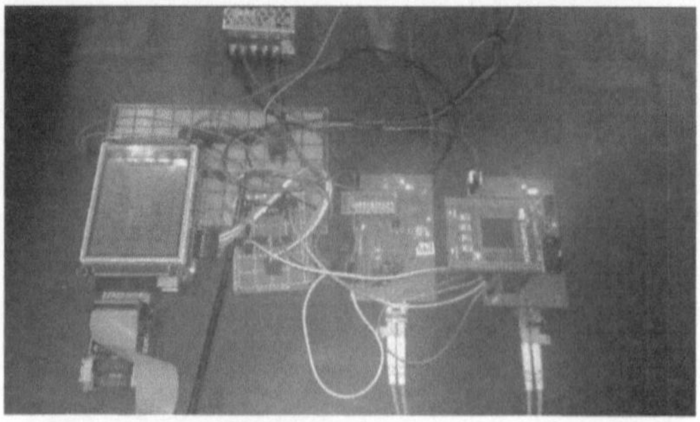

Figure 12. System global figure.

Table 1. The parameters of OTM.

Co.	Type	Data Rate Rate	WVL	DDM
Cisco	glc-sx-mmd	1250 Mbps	850 nm	YES
Huawei	esfp-ge-sx-mm-850	1250 Mbps	850 nm	YES
Nokoxin	sfp-sx-mm850	1250 Mbps	850 nm	NO

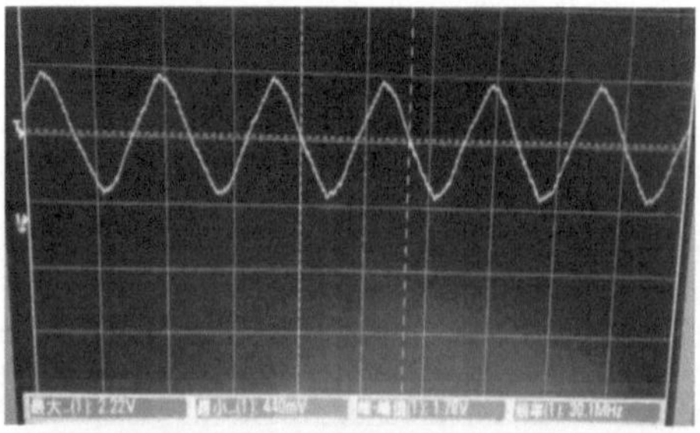

(a)

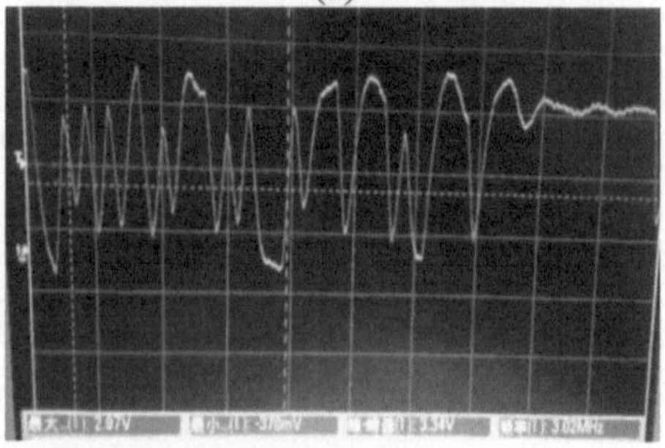

(b)

Figure 13. The SerDes output waveform in oscilloscope, (a) SerDes output when enable the OTM; (b) SerDes output when disable the OTM.

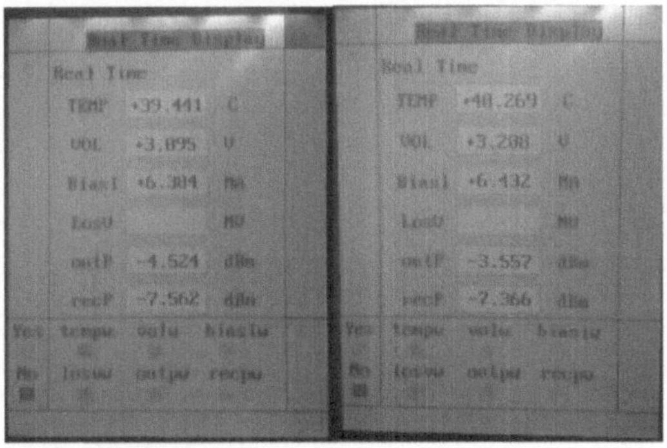

Figure 14. Monitoring values in TFT.

of LOS which is defined by SFF-8472 is extended. As long as the monitor values exceed the threshold values, it will generate an alarm or warning in the specified place. More than one parameters of OTM: temperature (TEMP), supply voltage (VOL), bias current (BiasI), tx_power (outP) and rx_power (recP) can be clearly displayed in TFT. Because of MSU, the system supports multiple modules at the same time, rather than to exchange OTM one after another to access into the system to obtain its monitoring values.

7 CONCLUSIONS

The multiple-channel and intelligence digital diagnostic system for SFP optical transceiver is presented in the paper. It is used to find the wrong point in optic fiber link, and it makes the communication more reliable as well as the data entry to OTM is more efficient. The newly multi-machine communication structure can be used to meet the requirements in the other system development.

ACKNOWLEDGMENTS

In this paper, the research is sponsored by the Nature Science Foundation of China (Project No. 61107025) and the Key Innovation Team Project of Zhejiang Province (Project No. 2010R50010).

REFERENCES

Hu Qinghong & Wu Youyu, 2013. *Research and Design of the Test Platform for SFP Optical Transceiver Module.* CNKI: China.

IIC *Bus Specification*, Version 2.1.2000.

Lei Yun & Qin Zhiguang, 2012. *Development and Application of the New Generation PON OLT/ONU.* CNKI: China.

Li Yang & Zhang Jinsong, 2008. *Research and Realization of Digital Diagnostics of SFP Optical Module.* CNKI: China.

SFF-8472 *Specification for Diagnostic Monitoring Interface for Optical Transceivers Rev 11.0.* 2010. SFF Committee.

Su Youzhang & Liu Shunkui & Zhou Jianyang, 2009. *Design and Implementation of SFP Optical Transceiver with Digital Diagnose Monitor Function.* CNKI: China.

Xia Xingxing & Ji Yuefeng, 2011. *Research on Performance Testing System of Optical Transceiver in Optical Access Network.* CNKI: China.

Yuan Rong, 2011. *Optical Fiber Communications Technology*, China Machine Press: China.

Yu Zheng, 2013. *The signal integrity revelation*, China Machine Press: China.

Zhou Jinlin & Zhang Yangan, 2009. *The Application and Realization of The Digital Diagnostic Monitoring Function for SFP Optical Transceiver Module.* Computing and Telecommunications: China.

Industrial Engineering and Management Science – Lee (Ed)
© 2015 Taylor & Francis Group, London, ISBN 978-1-138-02647-6

Application of digitization in innovation and preservation of religious art

LanYing Liu
Faculty of Art, Hubei University for Nationalities, Enshi, China

ABSTRACT: To restore and protect the dying religious art, the key point is digitized innovation and preservation of religious paintings and frescos by photographing or film shooting techniques as well as post-processing of database rebuilding. It is necessary to protect religious art. However, the constant digitized innovation is a prerequisite to explore the digitizing protection approaches, to present the art of Catholic church, provide maintaining, actualize its artistic value and to cope with the challenges in the process of material collection. Consequently, it has been proved that the digital technology involved in protecting religious art is a feasible and innovative method.

Keywords: Digitization; religious art; painting; fresco; preservation innovation

1 INTRODUCTION

Digital presentation and preservation of cultural heritage is emerging and developed gradually under the condition that the digital technique has been widely spread and reached its peak. They gradually come to the attention of most people who are involved in this field. The application of the advanced digital technology to protecting the valuable and rare cultural heritage is not only the good news for culture heritage protection workers, but also a demonstration of how the modern advanced technological innovation has been implemented[1]. The religious art includes traditional paintings, frescos, sacrificial utensils such as the Priests' vestments (using during the religious worships), Tabernacle, Eucharist Light, and chalices and so on. They are the holy, exquisite, simple and distinctive Catholic arts introduced to China by missionaries from Western countries as the sign of western religious civilization. Generally, those paintings and frescos in the Church have higher aesthetic value. In addition, they symbolize the golden age of the Catholic civilization in history. However, unfortunately some of Catholic paintings have been severely damaged due to historical event like the Cultural Revolution in 1960s.Some of them have fallen off or faded naturally as time goes by, especially the frescos. This thesis aims at restoring the religious art, such as frescos in the Catholic Church, sacrificial utensils and sacred images through textual arching, photographing, and video recording. The data collected through the technological means mentioned above, emphasize the integrity of unitary materials. But the link between various kinds of materials is sort of disconnected, especially it seems difficult to fully present the relations between all kinds of different materials while doing in-depth study. In the meantime, the paintings and frescos are still undergoing a series of changes due to the territorial differences and the extent of protection diversity. Thus, in view of protection and inheritance of religious art, more should be focused on working out the application strategies of digitization innovation and protection of Catholic Church art. At the same time, the data collection technology based on virtual reproduction has also been centered on all kinds of archaeological sites, museum of cultural relics, and ancient architecture of digital reconstruction projects and suchlike. For example, Dongming Lu, a Professor of the School of Computer Science at Zhejiang University, who presided the projects of "Archaeological Sites: Information Collection and Reconstruction Techniques" and "Digital Dunhuang"—a project invested by National Development and Reform Committee.[1] These projects put particular emphasis on recording the external information of the material objects, such as the materials, patterns and styles and so on. This thesis works on combining the traditional technique means and digitization techniques. The approach is to implement the integrated digital techniques, which consist of two parts, namely, the material collection and database building. They made an all-round information acquisition toward the cultural space of religious art. In addition, they proceeded with the database construction based on the collected information in order to prepare for the further presentation and research projects.

2 SCHEME OF COLLECTIONS

2.1 *Challenge*

The use of technology in cultural heritage faces a number of challenges and the precariousness of the objects manipulated during digitization. A number of innovation means of digitization, archiving and displaying have been developed. Collecting materials is the basic measure to protection of religious art. Religious art has the following distinguishing features: aside from displaying of frescos, it is easy to be collected. Besides, there are some utensils used in the religious rite which are visible, such as Tabernacle, Eucharist Light. Some other items only appeared during special religious worship. Since the altar area is the sacred place and it is off-limits to Non-Christians, the design in sacred area is inaccessible for collectors. Therefore, the interior structure form and artistic style of the alter area is not easy to catch. One solution is that those paintings, frescoes and religious utensils can be digitized as both 2D images and 3D spatial replica for different research and religious purposes. As for the contemporary situation, there is a lack of records of contemporary religious art status. And secularization directly influence the development of religious art and its regional flowing deformation. It results in the new challenges for rebuilding the religious art. In order to collect materials easily, the contents of religious art collection can be divided into three categories, namely: textual materials, fresco photographing, and on-site filming. This is virtually a form of "digital preservation". It is the process of bringing new life to religious art itself.

2.2 *Collecting stage*

2.2.1 *Textual materials*

The first step is to set up paper archives as the digital database of all manuscripts. The digitization of paper archive can be processed by scanning or digital camera photographing approach. While doing digital scanning, usually the mode is set at "Black and White Two-value". However one may use "grey scale" mode in case that the text is blurred. Generally, the scan resolution should reach 20 Odp, the images being scanned should be saved as TIF format. A minimum shooting degree of digital camera should be 3 million pixels, and the images should be saved as JPF format, and its zoom ratio should be 100%. Another tip is to make sure the frame is filled with the subject while shooting.

The following stage is to establish photography archives, to digitize the photographs with scanning. Scanning mode for color photographs should be set at "millions of color" mode, while "grey scale" mode for black and white photographs. As for those photographs of 5 inches (125 cm × 85 cm), the optical resolution of the scanner should reach 60 Odpi. For example, the photo to be scanned is a 1/N of the 5 inches photo, and then the scanner's optical resolution should be the N times of the 5 inches photo; After being scanned, the images should be saved as TIF format, and finally employ the LZW compression to those images. Last step is 3-D scanning which is mainly used to scan the object space profile and structure as well as its color, in order to obtain the spatial coordinates of object's surface. Its significance rests with being able to transform the physical three-dimensional information to digital signals. Using 3-D scanner to scan the fresco, paintings and sacrificial utensils, one can easily obtain its 3-D data. These datum can be connected to the CAD/CAM software interface directly. And in the CAD system, one can adjust, amend those datum freely, and also send them to machining center or rapid prototyping equipment for manufacturing. Thus it greatly shortens working hours.

2.2.2 *Fresco photographing*

a. Using location classification approach. Classify the frescos into vertical and lateral two categories according to their locations, and also measure them. The measurement result should determine its shooting plan by resolution ratio calculation method. It means that when the resolution ratio reaches X pixels per mm, then the coverage area of each photo should be (4/X)(3.2/X) meter. While the fresco's size is W m * H m, then, this fresco needs shooting of (H*W/3.2 rows, W*X/4 columns). As for each photo, it should be recorded as follows in the table: location, serial number, card number, camera, shutter, camera lens, aperture, the vertical distance above the ground and the lateral distance from the wall, light and daylight. While shooting with a whiteboard, the direction of flashlight needs adjusting. It means putting a piece of whiteboard on the left, right and the top of the camera's frame and shoot one for each in different directions. It needs to adjust the direction of flashlight on the base of the shooting results. After shooting and back up the photographs, employ the computer processing software to do image mosaic processing. Selecting the correct lights placement is an important step in the PTM acquisition of large objects. In general, we do not have the possibility to use a physical dome to illuminate the object. Instead, we will have to manually place the light in different positions, forming a "virtual" illumination dome. The size of this illumination dome and its light distribution will depend on the size of the target object and on the number of light directions we want to use. To simplify the light placements we developed a specific software tool that helps us to plan the positioning of the lights. The tool usage is quite simple. The scene setup is generated as the user inputs the size of the object to be acquired, its height from the ground and the distance of the camera. Objects in the scene are scaled according to user specifications. The camera is pointed towards the center of the object.
b. Using a Canon Digital Camera, a 1000 W halogen floodlight, a tripod and a boom stand. Only one light plains and the parallel-meridian placement of lights are used. With these configuration, one need to set the height and direction of the light only once for each level of height. The time needed to position the light was minimized by the acquisition planning just described and by some references placed on the floor. The shooting group fastened the acquisition using a printed scheme of the angle directions and

a plumb line attached to the light to facilitate the positioning.

The acquisition steps can be summarized as:

- Take the measures of the object, find the center of it and its height from the ground.
- Use these data, generate the "virtual dome" and put the reference marks following the output of the PTM planner.
- Position the digital camera on the tripod, and measure aperture and shutter speed under the illumination of the central light, then keep these values fixed for all the photos, in order to have a constant exposure.
- For each level of height, set the height and the direction of the light, then put it on each reference mark related to the level, and take the photo. Use image synthesis, and then, the photographs needs to be transferred into computer and using photoshop software to composite the images into an integral high-definition image.

2.2.3 *On-site filming*

Colleting materials mainly depends on photography and video graphy. Selecting equipment should be professional, accurate and portable. Both fixed camera set and mobile camera set are needed during shooting process. The former is used during Catholic Eucharistic celebration. The latter is for collecting frescos inside the church and the sacred images hang on the wall and so on outside the ceremony. On-site filming mainly aims at recording or filming the religious rites itself as well as obtaining the oral recording through exclusive interview. The recording pen should be prepared during the interview, and digitizing should choose 44.1 kHz as sampling criteria. After digitizing, save them in WAV or MP3 format. The data transfer rate of digitalization should not be less than 4 Mb/s, and should employ the MPEG2 compression storage standard. Since the image data needs post-editing, some information may get lost. To solve the problem, utilizing panoramic imaging technology to supplement the ceremony procedure becomes the issue of great importance. It should be recorded through micro-lens and be completed by displaying in full natural color of integral imaging techniques.

3 DATABASE BUILDING

First of all, to preserve the precious religious art consistently and effectively, how to establish the resource database about religious topic. Besides, how to reasonably apply the principle of computer programming, digitizing technique on collection, editing and processing, storage, demonstration and regeneration of the religious art. These are the issues that the Church protection workers face nowadays. First of all, basic protection work on those resources with relatively typical examples and incomplete yet valuable are still in the process. And the contents collected are incomplete. Moreover, to meet the needs of art researchers and the broad masses of believers, it will be better to choose those frescos demand simple technique. They can be completed in a short term to display instantly. In sum, improving the hardware requirement should highlight the environment where the open information can be created. Hardware build should be prospective. Its objective is to establish an advanced computer system and flexible structured cabling system, so that it can leave room for digitalization construction development. At the same time, we should adopt the system engineering which should be supported by advanced technology, and application of the information technology integration and the theory of digitalized antique resource database construction. The work aims at Church art and digital document resource database application background, employs the mode of client –server. It means applying the client, database server and object server to constitute the information transfer core structure. The digital resource database should be related to digital technology, super-large-scale database technology, network technique, multimedia information processing technology, database and analysis processing technique, information extraction technology, content based retrieval technique, and natural language understanding technology and so on and so forth comprehensive treatment, in order to carry out target data transfer in the resource database.

After that, the construction and development of digital document resources depends on high-quality talent. Human resource here means those staff who are professional in digitalizing cultural relic resources and the subsequent reserve of talents. This is the key to building the digital cultural resources. The ultimate goal for setting up the religious Church art data bank is to build a digital museum, apply the digital technology to preservation, storage and administration of religious cultural space related information. In the meantime, it can provide the clients with religious art information exhibition and research through the internet. The priority of database building is classification of the collected information and information based storage, database management can link up individual information effectively. By doing so, it can avoid the singleness and monotonicity of information. Database building must follow the principle of standardization and normalization. The software, database structure, and classification, used in the resources database building, and definition of graph-text field as well as the format of data entry and suchlike, must be standardized. All these factors contribute to the formation of comprehensive discipline system of database and provide people with information assurance facility. It is necessary to create the data catalogue according to the features of popular religious rites and classification coding system. This can be classified as four information retrieval categories, namely, text file, photographs bank, audio-visual record, animating and virtual class as well. It is also urgent to build index for a great variety of resources information for the purpose of facilitating the customer's researching.

4 CONCLUSION

Digitization of cultural relics is the foundation of computer-assisted preservation of cultural relics. The arrival of the digital age promises a solution to the problem of access. It also coincides with the rising demand from the general public as well as the scientific community of more immediate, more expansive and more detailed access to religious arts. As a result, despite the effort devoted in digital or on line preservation of cultural heritage, the general public is yet to be able to navigate them virtually in a coherent and educative way. However, despite the fact that it is impossible to include all materials in innovation and preservation of religious art, the ongoing collecting and compiling of the integrated works will be displayed to public successively, to employ the network technology gradually, as well as to achieve the extensive knowledge and acquire information communication. Therefore, the database building can become an integrated, systematic and comprehensive archives of religious art. It can fully display the permanence, diversity, systematicness, authenticity of religious art. The digital exhibition of religious art offers a valuable experience and opportunity to appreciate the interpretation of religious paintings and frescoes. In conclusion, whatever technology being used and wherever the art to be presented, to make the art information infinite, sharable and renewable is the everlasting goal of cultural relic preservation.

ACKNOWLEDGEMENT

This paper is one of the humanities and social science research projects supported by the Hubei Educational Commission of Hubei Province. Serial number: 2012jytq276.

REFERENCES

Eurographics 2008/M.Roussou and J Leigh

Jia Xieqing, Wang Jue, "Application of Digital Techniques in Inheritance and Innovation Domain of Chinese Cultural Heritage" (*Modern Media — Journal of Communication University of China*) Volume II, 2012

Lin Mei, "Introduction to Digitalization of Maijishan Grottoes Cultural Relics Resources" *Silk Roads*, Volume 22, 2010. (In Chinese)

Pan Yunhe, Lu Dongming, Chen Ren, "Virtual Rebuilding of Dunhuang Grottoes and Its Frescos Restoration Emulation" *ActaGeodaetica et* (In Chinese)

Roy S. Berns and Franziska S. Frey et al, Digital capture of cultural Heritage: Benchmarking American Museum. Practices and Defining Future Needs (Rochester, NY. 2005) p.57

Wang Ping, Pi Jiezheng, Hao Chunyun, Lu Weiguo, Yuan Xiaopeng, "Project Design on Digitalizing Dunhuang Grottoes", "*New Technology of Library and Information Service*" Nov., 2006, Volume 143, pp. 21–25. (In Chinese)

Industrial Engineering and Management Science – Lee (Ed)
© 2015 Taylor & Francis Group, London, ISBN 978-1-138-02647-6

Mesh figure evaluation method of hydrodynamic torque converter performance

A.L. Wang & W.G. Liu
School of Mechanic Engineering of Tongji University, Shanghai, China

ABSTRACT: Single or multiple weighting methods are difficult to evaluate hydrodynamic torque converter comprehensive performance. According to the problem, this paper proposes the mesh figure evaluation method of hydrodynamic torque converter Performance. It uses the launch torque ratio, the maximum efficiency, the impeller launch torque coefficient and the high efficiency area of hydrodynamic torque converters four main performance evaluation indicator. Besides, it also uses the maximum area of mesh figure as the optimized evaluation. In the practical study, we use the blade counts as the variable of optimization design. Then we use orthogonal CFD numerical simulation to analyze experiments. Furthermore, we use it to figure out how impeller, turbine and stator blade counts affect different performance indicators of hydrodynamic torque converter. The optimized blade counts results in better comprehensive performance of hydrodynamic torque converter. As a consequence, the phenomenon verifies the effectiveness of using mesh figure method to evaluate the designed performances of hydrodynamic torque converter.

1 INTRODUCTION

Hydrodynamic torque converter which realizes energy and power transmission function is an important part of transmission system of automobile and construction machinery. There are a lot of evaluation indicators for the performance of hydrodynamic torque converter, for example, launch torque ratio, the maximum efficiency, impeller launch torque coefficient, high efficiency area and so on (Zhu J.C., Liu Z. 2013). Most of the current studies on how to accurately evaluate hydrodynamic torque converter performance just stay in the evaluations method with single or multiple weights. It actually reflects the users' demand from different aspects.

Meanwhile, the importance of these indicators is not the same. That is to say, the results of Hydrodynamic Torque Converter construction optimization with this method are difficult to reflect the real performance of hydrodynamic torque converter. Mesh figure method was used in the evaluation of the stability of the vehicle by Sun Li (Sun L. & Liu Y.C. 2012). Meanwhile, based on mesh figure method, Chang Lv (Chang L. 2012) proposed the evaluation method for the vehicle with different engines and hydrodynamic torque converters. Mesh figure evaluation method avoids the manual distribution of the weight for each evaluation indicator.

Based on average stream theory, the optimization of hydrodynamic torque converter blade counts needs a lot of experiments and empirical formulas must be performed (Fan C.S. 2007). With the development and application of CFD and computer technology, using computer simulation instead of performing a large counts of experiments can greatly reduce the design cost and accelerate the optimization design process (Yan J. & He R. 2009). At present, most studies combined experiments with computer simulation, through a small counts of experiments calibrates the simulation results. The calibrated computer simulations results could replace the experiments, which are used in optimization of torque converter.

Therefore, this paper proposed the mesh figure evaluation method to evaluate the performances of hydrodynamic torque converter with different blade counts. It also uses the results of CFD numerical analysis combined with orthogonal experiments (Xu Z.A. & Wang T.B. 2002, Geniehi T. 1990). The experiments focused on study the influence of blade counts of impeller, turbine and stator in on different performance indicators (Liu C. & Pan X. 2012). After the optimization of blade counts, the performances of hydrodynamic torque converter are improved significantly. It verifies the effectiveness of mesh figure method which evaluates the designed performances of hydrodynamic torque converter.

2 ORTHOGONAL EXPERIMENT

2.1 *Factors and levels of orthogonal experiments design method*

In orthogonal experiments, the orthogonal table is used to arrange experiments. According to the characteristics of orthogonal table, the results of experiments are calculated and analyzed, in order to find a better process of test program (Wei Z.G. & Liu X.W. 1990).

Reasons or elements which may affect results of experiments are called factors. Factors in different

Table 1. Levels of orthogonal experiments design.

	A impeller	B turbine	C stator
Level 1	16	15	7
Level 2	20	19	10
Level 3	24	23	13
Level 4	28	27	16

Table 2. The orthogonal experiment of hydrodynamic torque converter blade.

No.	A	B	C	No.	A	B	C
1	1	1	1	9	3	1	3
2	1	2	2	10	3	2	4
3	1	3	3	11	3	3	1
4	1	4	4	12	3	4	2
5	2	1	2	13	4	1	4
6	2	2	1	14	4	2	3
7	2	3	4	15	4	3	2
8	2	4	3	16	4	4	1

states and conditions in the experiment can affect the results of experiments, these states and conditions of factors are called levels.

The design of orthogonal experiments of hydrodynamic torque converter blade counts has three factors: impeller blade counts, turbine blade counts, stator blade counts, which are denoted by A, B, C. In this paper the effective diameter of researched hydrodynamic torque converter is 355 mm. The initial impeller, turbine and stator blade counts are respectively 20, 18 and 14. The impeller, turbine and stator blade counts are divided into four levels, respectively 1, 2, 3, 4, as shown in table 1.

2.2 *The orthogonal design of hydrodynamic torque converter blade counts*

According to Table 1, the orthogonal experiment of hydrodynamic torque converter blade counts has shown in Table 2.

2.3 *The orthogonal experiment results of hydrodynamic torque converter blade counts*

Results of Hydrodynamic Torque Converter CFD numerical analysis has shown in Table 3. It can be seen from Table 3, blade counts of impeller, turbine and stator may affect the Launch torque ratio, peak efficiency, launch torque coefficient of impeller and high efficiency area. Now, through the main effects chart and Pareto charts, characterize the degree of Effect on different characteristics of the blade counts. Mainly the linear correlation degree, quadratic correlation degree and interaction effect degree are included.

Fig. 1 a and b are the main effects chart and Pareto chart of impeller, turbine and stator blade counts's peak efficiency. It's obvious that stator's blade counts have a great effect on peak efficiency, far more than the effect of turbine and impeller. Effect of stator's blade counts to maximum efficiency shows a distinct secondary correlation which is an approximate opening down parabola. effect of impeller blade's counts to peak efficiency shows an approximate single reduced relationship, and the effect is far less than stator's. Effect of turbine blade's counts to peak efficiency shows an approximate opening up parabola relationship. Also, the effect is far less than stator's.

Table 3. Results of CFD numerical analysis.

No.	K_0	η	λ_B	G
1	2.1210	0.8272	3.1544	1.7515
2	2.2708	0.8488	3.1503	1.9247
3	2.3242	0.8580	3.2014	1.9825
4	2.3464	0.8514	3.1909	1.9584
5	2.1607	0.8440	3.1385	1.5258
6	2.2149	0.8427	3.1511	1.8739
7	2.3091	0.8637	3.4542	2.0073
8	1.9704	0.8190	3.0974	1.6779
9	1.5166	0.7425	2.6013	1.1912
10	1.6936	0.7731	2.9659	1.3822
11	2.2013	0.8496	3.5815	1.8898
12	1.9867	0.8181	3.4575	1.6808
13	1.5542	0.7399	2.4636	1.1275
14	1.7327	0.7838	2.9839	1.4424
15	1.8897	0.8029	3.2026	1.5926
16	1.5860	0.7355	3.2688	1.0000

Fig. 2 a and b are the main effects chart and Pareto chart of impeller, turbine and stator blade counts launch torque coefficient of impeller. It is obvious that stator blade counts have a great Effect on launch torque coefficient of impeller, far more than the effect of turbine and impeller. Effect of stator blade counts to launch torque coefficient of impeller shows a distinct secondary correlation which is an approximate opening down parabola. Effect of turbine blade counts to launch torque coefficient of impeller shows an approximate opening up parabola relationship. Also, the Effect is far less than stator.

Fig. 3, a and b respectively for main effects chart and Pareto chart of high efficient area of the impeller, turbine and stator blade counts. It is not difficult to see as follows:

1) The influence on the high efficient area of the stator blade counts is far more than the influence of other wheels blade counts;
2) The correlation between stator blade counts and high efficient area is an obvious two function of parabola with downward opening:
3) The influence on the high efficient area of the impeller blade counts is an approximately monotone decreasing correlation, but the influence is far less than stator;
4) The influence on the high efficient area of the turbine blade counts is a two function of parabola with upward opening, and the influence is far less than the stator as well.

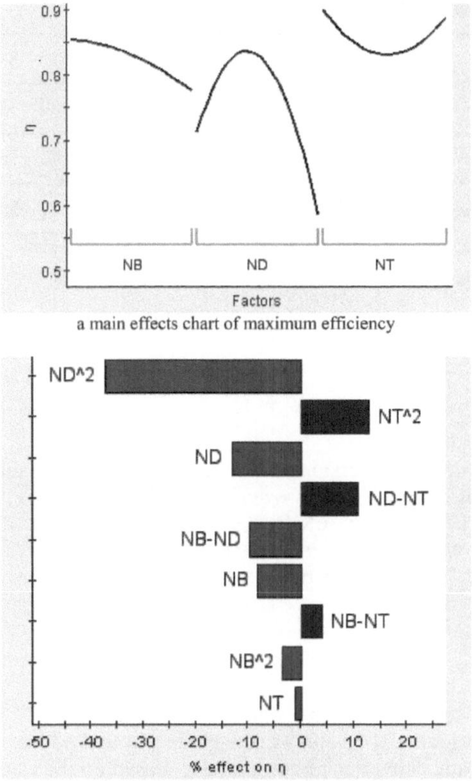

a main effects chart of maximum efficiency

b Pareto chart of launch torque ratio

Figure 1. Effect of blade counts on launch torque ratio.

3 MESH FIGURE EVALUATION OF PERFORMANCE

There is a certain number of performance indicators of hydrodynamic torque converter, which mainly include launch torque ratio, maximum efficiency, impeller launch torque coefficient and high efficient area, etc. Generally, according to experience, performance evaluation of hydrodynamic torque converter mainly aimed at optimizing an indicator by neglecting some secondary performance indicator. Or it aimed at properly distributing the weight of the indicators by taking these performance indicators into consideration.

However, these methods are hard to reflect the comprehensive performance of hydrodynamic torque converter, often guaranteeing only one good performance indicator. Hydrodynamic torque converter performance evaluation based on the network map can be very good considering each performance indicator of the hydrodynamic torque converter. Thus it can optimize and get the optimal combination of comprehensive performance of hydrodynamic torque converter.

3.1 *The establishment of Mesh figure evaluation of performance*

Hydrodynamic torque converter performance evaluation is a multi-objective optimization problem.

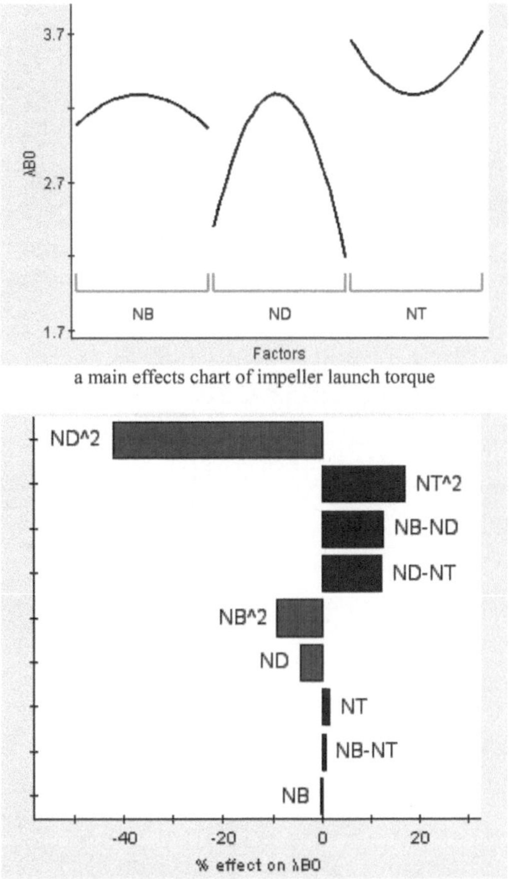

a main effects chart of impeller launch torque

b Pareto chart of impeller launch torque coefficient

Figure 2. Effect of blade counts on impeller launch torque coefficient.

Sub-goals value should be quantifying and calculating. After normalized treatment, the value obtained would be treated as the performance indicator value. The calculation of the value is shown in Fig. 4.

Establish mesh circle evaluation of performances as Fig. 4, O point is the center of the circle. $\overline{OA} = a_1 = K_0$, $\overline{OB} = a_2 = \eta$, $\overline{OC} = a_3 = \lambda_B$, $\overline{OD} = a_4 = G$

where:

K_0–Launch torque ratio;

η–Maximum efficiency;

λ_B–Impeller launch torque coefficient;

G–High efficient area.

$\beta_1, \beta_2, \beta_3, \beta_4$ are the angles between a_1, a_2, a_3, a_4. The angles are calculated as the below formulas:

$$\beta_i = \frac{a_i + a_{i+1}}{\sum_{j=1}^{4} a_j} \times 180 \qquad (i = 1,2,3;\ j = 1,2,3,4,)$$

$$\beta_i = \frac{a_4 + a_1}{\sum_{j=1}^{4} a_j} \times 180 \qquad (i = 4;\ j = 1,2,3,4,)$$

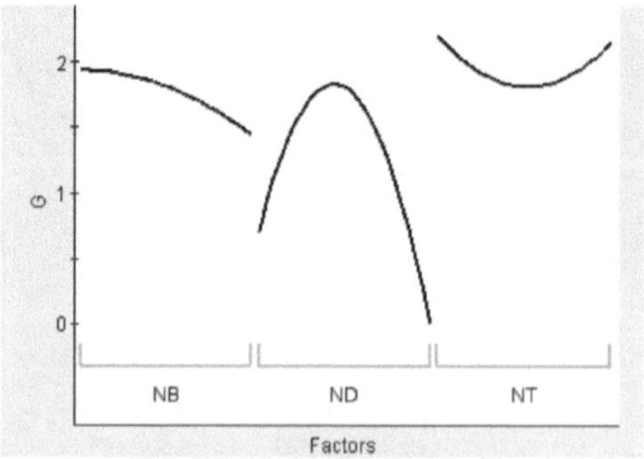

a main effects chart of high efficiency area

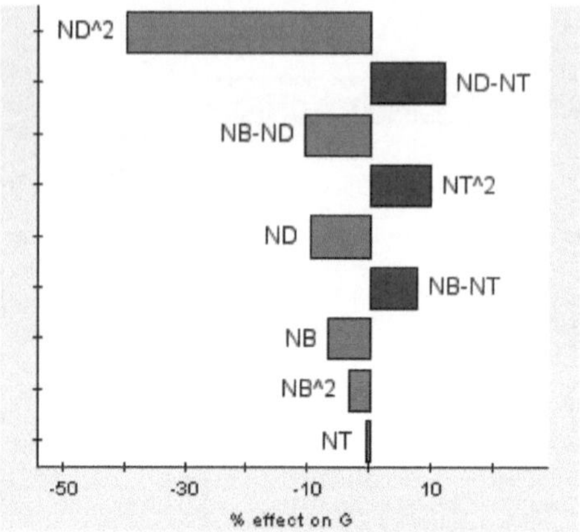

b Pareto chart of high efficiency area

Figure 3. Effect of blades counts on high efficiency area.

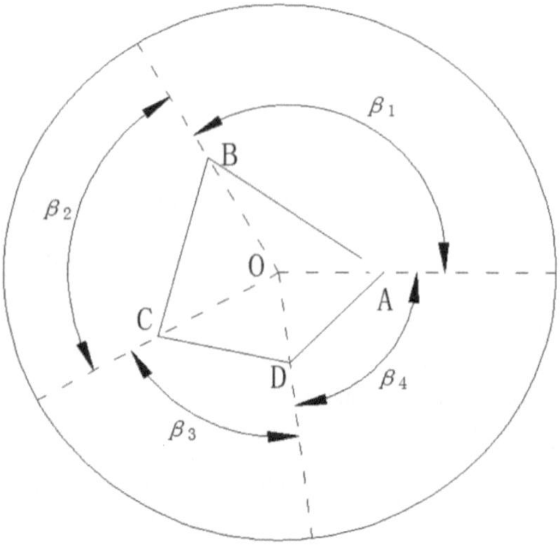

Figure 4. Mesh figure evaluation of performances.

3.2 *Design variables and objective function*

The design variables of hydrodynamic torque converter blade counts optimization are impeller blade counts, turbine blade counts and stator blade counts, namely:

$$X = [N_B \quad N_T \quad N_D]$$

In the expression above, X is the design variables matrix, N_B, N_T, N_D are design variables.

Each set value of N_B, N_T, N_D can be calculated out a set value of a_1, a_2, a_3, a_4, and then calculate the area of the quadrilateral ABCD, expressed as S_{ABCD}, namely:

$$\begin{aligned} S_{ABCD} &= S_{\Delta OAB} + S_{\Delta OBC} + S_{\Delta OCD} + S_{\Delta ODA} \\ &= 0.5a_1a_2\sin\beta_1 + 0.5a_2a_3\sin\beta_2 + 0.5a_3a_4\sin\beta_3 + 0.5a_4a_1\sin\beta_4 \end{aligned}$$

According to the significance of a1, a2, a3, a4, the higher the value is, the better the hydrodynamic torque converter performance is. Based on the established method of mesh figure evaluation, the bigger the S_{ABCD} is, the better the hydrodynamic torque converter performance is. S_{ABCD} actually reflects the comprehensive performance of hydrodynamic torque converter. Establish the objective function:

$$f(X) = S_{ABCD} \rightarrow \max$$

Table 4. S_{ABCD}.

No.	1	2	3	4	5	6	7	8
S_{ABCD}	6.53	7.24	7.55	7.50	6.08	7.01	7.89	6.04
No.	9	10	11	12	13	14	15	16
S_{ABCD}	3.79	4.74	7.49	6.39	3.62	4.97	5.77	3.87

3.3 *Results analysis and optimization*

According to table 3, S_{ABCD} can be calculated and shown in table 4.

In Fig. 5, we can seen a and b respectively for main effects chart and Pareto chart of high efficient area of the impeller, turbine and stator blade counts. It is not difficult to see as follows:

1) The influence on the high efficient area of the stator blade counts is far more than the influence of other wheels blade counts;
2) The correlation between stator blade counts and high efficient area is an obvious two function of parabola with downward opening;
3) The influence on the high efficient area of the impeller blade counts is an approximately monotone decreasing correlation, but the influence is far less than stator;
4) The influence on the high efficient area of the turbine blade counts is a two function of parabola with upward opening, and the influence is far less than the stator as well.

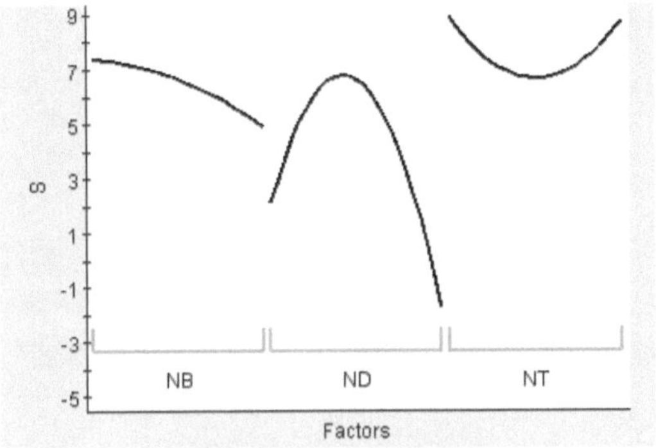

a main effects chart of SABCD

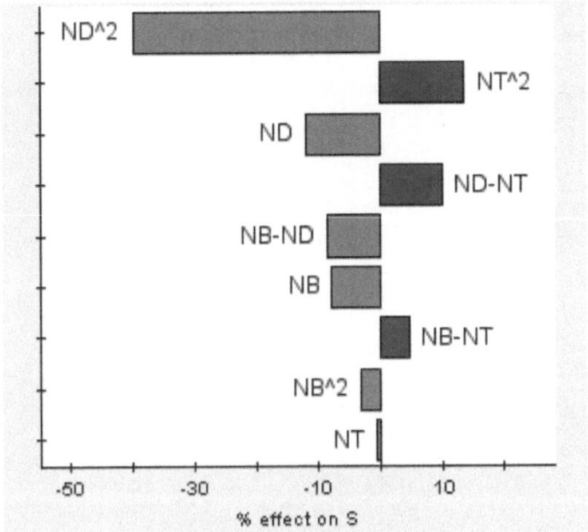

b Pareto chart of SABCD

Figure 5. Effect of blades counts on SABCD.

Table 5. Performance comparison before and after optimization.

	K_0	η	λ_B	G	S_{ABCD}
before	2.1983	0.8388	2.8652	1.8291	6.5486
after	2.3091	0.8637	3.4542	2.0073	7.8919

Response surface methodology is used to seek the maximum of S_{ABCD} and optimize. The maximum of S_{ABCD} is 7.8919 when the impeller, turbine and stator blade counts are respective 20, 27, 13. The comparison between optimized performance and initial performance is shown in table 5.

The table 5 shows that performance evaluation indicators of the optimized hydrodynamic torque converter such as starting torque ratio, the maximum efficiency, impeller starting torque coefficient and high efficient area are improved. By comparing the optimized model with the initial one, turbine blade counts is increasing. Stator blade counts is decreasing, making the area of the network larger than the original one, which presents that the comprehensive performance of hydrodynamic torque converter is improved significantly.

4 CONCLUSIONS

(1) Effectively reduce the required experiments counts and achieve the purpose that saves the test time and cost through the method of orthogonal CFD numerical simulation and analysis experiments.
(2) Present the mesh figure evaluation method of Hydrodynamic torque converter performance that breaks the single or multiple weighting methods and achieves better results.
(3) Stator blade counts have a deeper influence of hydrodynamic torque converter performance than impeller and turbine. The relation between stator blade counts and the performances is mainly quadratic relation which is a downwards parabola relation.

REFERENCES

Chang L. 2012. Optimization of power matching on torque converter with diesel engine for wheel loader based on performance evaluation mesh figure. Transactions of the CSAE, 01, 50–54.

Fan C.S. 2007. Study on influence of blade counts to performances of hydrodynamic torque converter. Changchun: Jilin University.

Geniehi T. 1990. Principles of design of experiment. Beijing: Ordnance Industry Publishing House:9–12.

Liu C. & Pan X. 2012. Effect of Blade Number on Performance of Torque Converter and Its Optimization Based on DOE and Response Surface Methodology. Transactions of Beijing Institute of Technology, 07:689–693.

Liu Z. 2013. Research of Engine and Hydraulic Torque Converter Matching and Evaluation Criterion for Loader. Chang'an University,

Sun L. & Liu Y.C. 2012. An Improved Net map Method for Vehicle Control Stability Evaluation. Machinery Design and Manufacture, 06:64–65.

Xu Z.A. & Wang T.B. 2002. Brief Introduction to the Orthogonal Test Design. Sic/Tech Information Development and Economy, 05:148–150.

Yan J. & He R. 2009. Numerical simulation of hydrodynamic retarder with different blade counts. Journal of Jiangsu University, 30(1):27–31.

Wei Z.G. & Liu X.W. 1990. Zhou Sihong. Brief Introduction to the Orthogonal Test Design. Journal of Chemical Education, 06:17–21.

Zhu J.C. Torque converter design and calculation. Beijing: National Defense Industry Press

Section 2: Industrial development and industrial engineering

Industrial Engineering and Management Science – Lee (Ed)
© 2015 Taylor & Francis Group, London, ISBN 978-1-138-02647-6

Adaptive filtering queueing with approximate fairness

Jui-Pin Yang
Department of Information Technology and Communication, Shih-Chien University, Taiwan, R.O.C.

ABSTRACT: In this paper, we propose an Adaptive Filtering Queueing (AFQ) algorithm that achieves approximate fair bandwidth sharing among competing flows. AFQ employs a filtering level table that hierarchically filters arriving packets. Based on filtering levels of arriving packets and estimates of fair filtering levels, AFQ determines the exact treatment on arriving packets by simple packet dropping. Simulation results show that the fairness of AFQ outperforms RPQ and CSFQ, and much better than CHOKe and FIFO under a variety of traffic conditions.

1 INTRODUCTION

Random Early Detection (RED) effectively improves congested conditions by monitoring and controlling average queue size [1]. RED not only keeps queuing delay low but also retains high throughput because it prevents current connections from global synchronization. Unfortunately, RED has to cooperate with specific transport-layer protocol, namely TCP. Besides, RED is incapable of dealing with fair bandwidth sharing among flows. To improve RED's fairness while keeping simplicity, Choose and keep for responsive flows, choose and kill for unresponsive flows (CHOKe) was proposed [2]. When a packet arrives at a congested router, CHOKe randomly draws a residing packet and compares whether both packets come from the same flow. If they both belong to the same flow, both are discarded at the same time; otherwise, the arriving packet is admitted to enter the buffer according to a probability that depends on the degrees of congestion. XCHOKe is a revised version of the CHOKe, which detects possible malicious flows by storing the flow labels of CHOKe hits in a lookup table [3]. If a flow label has many hits, such flow has a higher chance to be identified as a malicious flow. Accordingly, its arriving packets are assigned with a higher dropping probability. However, XCHOKe has limited improvement on fairness while comparing with CHOKe and RED.

Considering feasibility and reasonable degree of fairness, Core-Stateless Fair Queueing (CSFQ) [4] and Rainbow Fair Queueing (RFQ) [5] are proposed. Routers in CSFQ are classified into edge and core. Edge routers need to maintain per-flow state, in other words, they have to estimate the flow rate and insert the state into corresponding packet headers. Base on the state, core routers simplifies the estimation of dropping probability on arriving packets. The idea of RFQ is similar to CSFQ. In general, they both may fail in fairness when the number of traverse links increases. The reason is that inaccurate flow state is amplified. Rotating Preference Queues (RPQ) is composed of a set of FIFO output queues that are dynamically rotated [6]. RPQ dispatches qualified arriving packets to adequate output queues based on packet distribution and preference. RPQ achieves significant fairness but it needs large buffer size. In another word, RPQ has higher implementation costs and queueing delay. Compared with RPQ, CSFQ, CHOKe and FIFO, AFQ achieves approximately perfect fairness.

2 ADAPTIVE FILTERING QUEUEING (AFQ)

Figure 1 shows that AFQ consists of four components including computation of aggressive traffic, fair filtering level estimation, a filtering level table and a packet dropping method. Also, we propose an alternative architecture to simplify the AFQ's implementation by integrating with the concept of CSFQ [5]. If the mean arriving rate of a flow is larger than max-min fair rate, such a flow is identified as an aggressive flow; otherwise, it is a non-aggressive flow. In addition, a flow id is defined as a pair of IP source-destination addresses that identifies different flows.

At each router, we use Equation (1) to estimate the aggregate traffic $\hat{A}$ passing through the packet dropping component in a time interval.

At each router, we use Equation (1) to estimate the aggregate traffic $\hat{A}$ passing through the packet dropping component in a time interval.

$$\hat{A}_{new} = k_a A_{now} + (1 + k_a)\hat{A}_{old} \quad (1)$$

$\hat{A}_{old}$ is the value before the updating of $\hat{A}_{new}$ after-a-time interval and $\hat{A}_{now}$ is the amount of aggregate traffic in current time interval. Furthermore, k_a is a

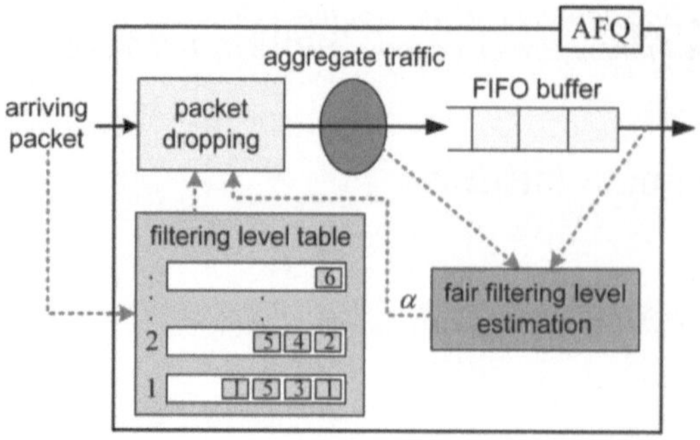

Figure 1. AFQ.

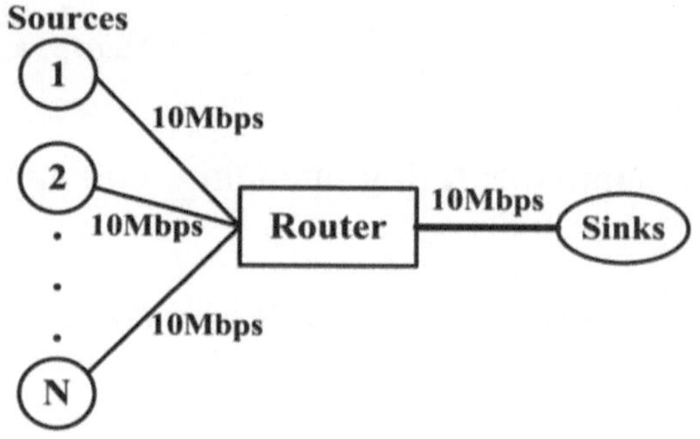

Figure 2. Network topology.

constant and T_d denotes the length of time interval. Next, we estimate the fair filtering level $\hat{a}$ by a linear relationship and this yields Equation (2). Similarly, $\hat{a}_{old}$ is the value before the updating of $\hat{a}_{new}$ after-a-time interval and denotes the link capacity.

$$\hat{a}_{new} = \hat{a}_{old} C T_d / \hat{A}_{new} \tag{2}$$

When a packet arrives, AFQ orderly compare its flow id with a randomly selected flow id from filtering level 1 to $\hat{a}_{new}$ filtering level table until there is a hit. The filtering level table consists of filtering levels which store flow ids only. If both have the same flow id (i.e. coming from the same flow), a hit happened. Each filtering level only stores flow ids not packets and this information is used to discriminate acceptability of arriving packets. In other words, multiple filtering levels work like a hierarchical filter that sifts qualified packets out. The discriminability of AFQ on arriving packets may be degraded because of traffic dynamics. For instance, $\hat{a}_{new}$ increases. In Equation (3), we re-adjust the number of flow id comparisons by k_b times the $\hat{a}_{new}$, denoted by $\hat{F}_{\max}$. Therefore, the range of flow id comparisons is altered from filtering level 1 to $\hat{F}_{\max}$

$$\hat{F}_{\max} = k_b \hat{a}_{new} \text{ where } k_b > 1 \tag{3}$$

If packet i has a hit at filtering level m_i, then we use Equation (4) to calculate filtering level v_i where its flow id is stored. There are two supplementary rules. First, if i packet encounters an empty filtering level at v_{emp}, then $v_i = v_{emp}$. Second, if there is no hit until $\hat{F}_{\max}$, then $v_i = \hat{F}_{\max}$.

$$v_i = \begin{cases} m_i - 1 & 2 < m_i \le \hat{F}_{\max} \\ 1 & 1 < m_i \le 2 \end{cases} \tag{4}$$

After determining the filtering level v_i, we design a circular replacement method to assign the location of flow id $index_i$ by Equation (5). The initial value of $index_i$ is of zero. Also, we assume that all filtering levels have the same length, denoted by L. By using the method, AFQ keeps the latest update of traffic status that is helpful to enhance fairness.

$$index_i = max(index_i + 1) mod(L+1), 1) \tag{5}$$

Finally, a simple packet dropping method is applied to determine the treatment on packet referring to Equation (6), where $prob_i$ denotes the dropping probability of packet i. Once packet i is qualified to be accepted, AFQ put this one on FIFO buffer; otherwise, it is discarded.

$$prob_i = \min(floor(a_{new}/m_i), 1) \tag{6}$$

3 SIMULATION RESULTS

We simulated the other four algorithms to compare their fairness with AFQ including RPQ, CSFQ, CHOKe and FIFO. According to simulation results, we find that AFQ achieves the best degree of fairness, significantly over others. We use a software simulator to perform all simulations [6]. The traffic types of generating packets for each case are described in respective figures. In Fig. 2, we consider a single congested link. Unless otherwise specified, we use the following configurations in all simulations. Each link capacity is of 10 Mbps and packet size is fixed at 1 kB. The buffer size for all algorithms is set to 256 kB. In addition, we neglect the propagation delay on each link. In AFQ, the initial setting of a_{new} is of 32, and the other parameters are set to $T_d = 200$ ms, $k_a = 0.8$ and $k_b = 1.5$. In RPQ, each output queue is set to 32 kB, and the other parameters are set as follows: $\Delta = 0.8$ ms, $a = 0.8$, $K_d = 200$ ms and $N = 129$. With respect to CSFQ, K and K_a are both set to 200 ms. CHOKe's parameters are set to the following values: $max_{th} = 120$ kB, $min_{th} = 40$ kB, $w_q = 0.002$ and $max_p = 0.02$. Finally, each simulation time is of 200 seconds.

We consider two cases where a single congested link is shared by 10 flows in Fig. 2. In the first case, all flows are indexed from 1 to 10 and the max-min fair share rate is 1.4 Mbps. Figure 3 shows the Normalized Bandwidth Ratio (NBR) of each flow [6]; CHOKe and FIFO both achieve limited fairness, with flows 1 to 5 getting bandwidth proportional to their mean arriving

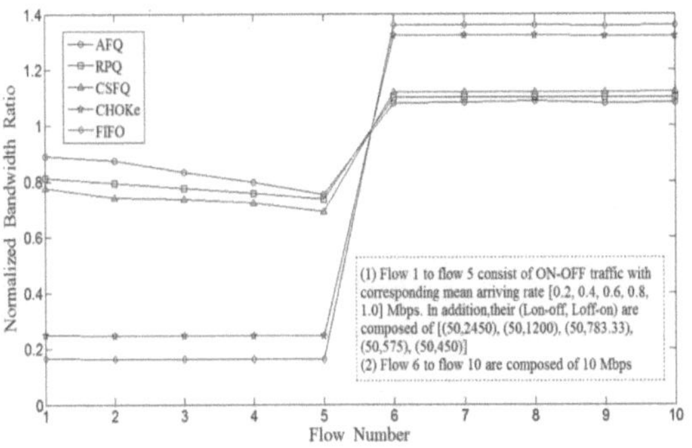

Figure 3. NBR achieved by each of ten flows sharing a bottleneck link, and flows 1 to 5 belong to non-aggressive flows and flows 6 to 10 belong to aggressive flows.

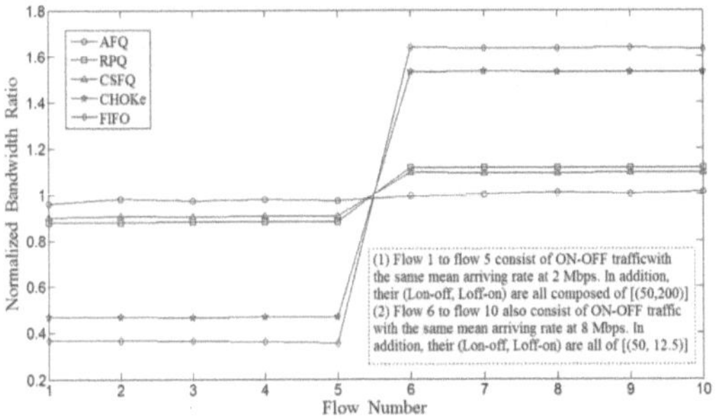

Figure 4. NBR achieved by each of ten flows sharing a bottleneck link, and all flows belong to aggressive flows.

rate, so their NBRs are close to a constant. Flows to 10 have the mean arriving rate, so they roughly fair share the bandwidth from flows 1 to 5. Consequently, their NBRs keep constant. AFQ is extremely effective in achieving fair bandwidth sharing while RPQ and CSFQ achieve a worse degree of fairness. The NBRs of flows 1 to 5 in AFQ, RPQ and CSFQ decrease while mean arriving rate approximates to max-min fair share rate.

In the second case we consider that all flows belong to aggressive flows. Also, the max-min fair share rate is 1 Mbps. Figure 4 shows the NBR of each flow. Compared with the first case, CHOKe and FIFO perform better because of less difference to incoming flow rate. The only algorithm that can most effectively achieve the perfect fairness is AFQ. RPQ and CSFQ have approaching fairness but they both are still worse than the AFQ. In a word, AFQ demonstrates the best fairness under different traffic conditions.

4 CONCLUSIONS

We present AFQ algorithm for achieving approximately fair bandwidth allocations. Each router performs simple packet dropping on arriving packets based on their filtering levels and estimates of the fair filtering levels. In addition, we also propose an alternative architecture to simplify comparisons while reducing memory consumption. We analyzed the fairness of AFQ, and four other algorithms under a variety of conditions. Simulations show that AFQ's fairness is superior to the RPQ and CSFQ, and much better than the CHOKe and FIFO.

ACKNOWLEDGEMENT

The author acknowledges the financial support from National Science Council in Taiwan, Republic of China under the grant numbers NSC-102-2221-E-158-001.

REFERENCES

[1] M. Shreedhar and G. Varghese, "Efficient fair queuing using deficit round-robin," IEEE/ACM Trans. Networking, vol. 4, no. 3, pp. 375–385, 1996.

[2] R. Pan, B. Prabhakar and K. Psounis, "CHOKe: a stateless active queue management scheme for approximating fair bandwidth allocation," in Proc. of IEEE INFOCOM, pp. 942–951, 2000.

[3] P. Chhabra, A. John, H. Saran and R. Shorey, "Controlling malicious sources at Internet gateways," in Proc. of IEEE ICC, pp. 1636–1640, 2003.

[4] I. Stoica, S. Shenker and H. Zhang, "Core-stateless fair queueing: a scalable architecture to approximate fair bandwidth allocations in high-speed networks," IEEE/ACM Trans. Networking, vol. 11, no. 1, pp. 33–46, 2003.

[5] Z. Cao, Z. Wang and E. Zegura, "Rainbow fair queueing: fair bandwidth sharing without per-flow state," in Proc. of IEEE INFOCOM, pp. 922–931, 2000.

[6] J. P. Yang, "Rotating preference queues: an efficient queue management scheme for fair bandwidth allocation," IEEE Communications Letters, vol. 17, no. 2, pp. 420–423. 2013.

Industrial Engineering and Management Science – Lee (Ed)
© 2015 Taylor & Francis Group, London, ISBN 978-1-138-02647-6

Determining characteristic joints during monotonous tasks with motion sensor

Shun Okamura & Yohei Tontani
Graduate School of Science and Engineering, Ritsumeikan University, Shiga, Japan

Yusuke Kajiwara & Hiromitsu Shimakawa
College of Information Science and Engineering, Ritsumeikan University, Shiga, Japan

ABSTRACT: Manufacturing industry features the line production system, in which workers repeat monotonous works. The monotonous work makes workers tired, and sometimes sleepy. Therefore, the monotonous work causes human errors. This research addresses the measurement of worker concentration with accelerometers attached to them. However, we do not know effective attachment points of accelerometers to be attached. The effective points should be worker joints showing apparent difference in movement at concentration time and non-concentration time. This paper refers an effective attachment point as to a characteristic joint. Since characteristic joints vary with workers or tasks, it is essential to determine characteristic joints. This paper proposes an automatic method to determine characteristic joints for a monotonous work. In this method, a vision-based motion sensor receives movement data of each joint of the worker. A machine learning method is applied to find worker joints apparent in their movement difference at concentration time and non-concentration time. One method extracts the most apparent worker joints and characteristic joints, reflecting the diversity of workers and tasks. Accelerometers attached to characteristic joints detects the movement of tired workers, which prevents the human error of monotonous work.

1 INTRODUCTION

Working sites in manufacturing industry are still experiencing many accidents. As measures against human errors, workers are encouraged to confirm by pointing and calling [1], take a regular rest, and so on. However, the decline of concentration prevents the confirmation itself. The duration of lasting concentration varies with each worker. It is difficult to address the problem with the measures taken in a fixed period. Even if a method is devised to increase concentration or to keep longer concentration, the decline in concentration should be detected for each worker. Detection of concentration decline for each worker would solve problems attributed to individual difference, which greatly reduce human errors. A method to detect a decline in concentration is indispensable to establish safe production lines with high productivity.

The authors have proposed a method [2] to detect a decline of worker concentration from his body movement. In the method, a worker wears two accelerometers on his head and right wrist. The method calculates the degree of worker concentration, measuring the stability in the movement of his head and right wrist. However, the method does not consider the effectiveness of attachment points of accelerometers other than the head and right wrist. Indeed, some workers show unusual movements in the two parts when they are out of concentration, but there might be workers who have attachment points of accelerometers showing their concentration more effectively than their heads and right wrists. It is necessary to determine effective attachment points of accelerometers for each worker. It provides criteria to take a rest at timing appropriate for each worker.

The method this paper proposes determines important joints depending on each worker and each task in monotonous tasks. This method uses a motion sensor based on the image processing. The motion sensor gets joint movement of a worker engaging in a monotonous task both with concentration and without. Utilizing the machine learning, we determine important joints from differences in movements during working under concentration and out of concentration. Since this method uses the motion sensor to gets joint movements of the whole body of workers, they do not have to wear any sensor which may interfere in their working. The machine learning toward joints of their whole bodies enables to find important joints systematically. An experiment has been conducted to show the effectiveness of the method, implementing an working environment for a monotonous task. This experiment focuses on the identification of important joints through movements of each joint of testees. As a result, we have succeeded in determining important joints for almost all testees. We have classified the testees, based on the important joints. It has revealed the commonality of important joints classifies workers into four types. The fact suggests that we can judge important joints only from observation of different movement in roughly partitioned body parts between working under concentration and one out of concentration.

2 BACKGROUND

2.1 *Problem of line production*

A line production system is one of major production systems in manufacturing industry. In the line production system, a worker has to repeat a single task toward lots moving on the conveyer belt, to make one product. In the system, heavy burden is imposed on each worker who has to repeat a single task monotonously for a long time. The burden makes the concentration of a worker decline, which leads to injuries or careless mistakes. To prevent the decline of concentration, workers are encouraged to take measures such as confirmation by pointing and calling [1], taking a regular rest, and so on. However, the duration of lasting concentration varies with each worker. It is difficult to address the problem with the measures taken in a fixed period. Even if a method is devised to increase concentration or to keep longer concentration, the decline in concentration should be detected for each worker. Real-time detection of a decline in concentration of each worker would prevent the worker from injuries and careless mistakes. A method to detect a decline in concentration is indispensable to establish safe production lines with high productivity.

2.2 *Base approach*

The authors have proposed a method [2] to detect a decline of worker concentration from his body movement. In the method, a worker is attached with two accelerometers on his head and right wrist. The method calculates the degree of worker concentration, measuring the movement of his head and right wrist. The method learns the movement of a worker of the concentration time in advance. Using the difference between the movement learning in advance and working actually detect worker concentration. The method detects the degree of worker concentration from the movement of the worker wearing two accelerometers. The method is useful because it provides measures to take a rest at timing appropriate for each worker. However, the method does not consider the effectiveness of attachment points of accelerometers other than the head and right wrist. There would be workers who have attachment points of accelerometers showing their concentration more effectively than their heads and right wrists. A method is indispensable to determine effective attachment points of accelerometers.

2.3 *Existing research*

Many works try to grasp mental states of persons with accelerometers attached to their bodies. The research [3] measures group behaviors of children by accelerometer attached to the waist of children. It aims at contributing to healthy growth of children via analyzing relationships of them.

Another research [4] measures the child movement looking back with accelerometers and cameras while childcare worker read picture books to many children. The measurement reveals the concentration of the children. Childcare workers are informed of their concentration, to make children interested in the reading of picture books. These researches do not consider what joints indicate the mental states of persons by their movement. There is no guarantee that the selected joints indicate the mental status best. This research proposes the method that determines the best joint for accelerometers to be attached.

3 DETERMINING CHARACTERISTIC WITH KINECT

3.1 *Overview*

The method this paper proposes determines characteristic joints depending on each worker and each tasks in monotonous tasks. This paper refers to characteristic joints as joins which indicate well the concentration of a worker when they are engaging in monotonous tasks. It is indispensable to measure the movements of joints in the whole body, because we do not know characteristic joints depending on each worker or each task. It is assumed that accelerometers attached to joints of the whole body measure the movements of joints in the whole body of a worker. However, it is a big burden for workers to measure movements of joints of the whole body with contact motion sensors such as accelometers. This method uses Kinect, which is a contactless motion sensor. It can measure joints of the whole body of a worker at the same time.

Fig. 1 shows the overview of the method. The Kinect sensor gets joint position of a worker engaging a monotonous task both with concentration and without. The Kinect sensor can determines characteristic joints effectively because the Kinect sensor gets joint position of the whole body of a worker without burdening workers. The identification of characteristic joints enables us to judge the concentration of workers through the observation of the movement of their characteristic joints during monotonous tasks for a long time. The judgment contributes to preventing human errors which brings serious injuries and tremendous loss caused by careless mistakes. This paper focuses on the identification of characteristic joints.

3.2 *Getting movement data with Kinect*

This method determines characteristic joints with a Kinect sensor. The Kinect sensor functions as the low cost contactless motion sensor that can measure joint position of the whole body of workers because the sensor grasps the skeleton of a worker measurement

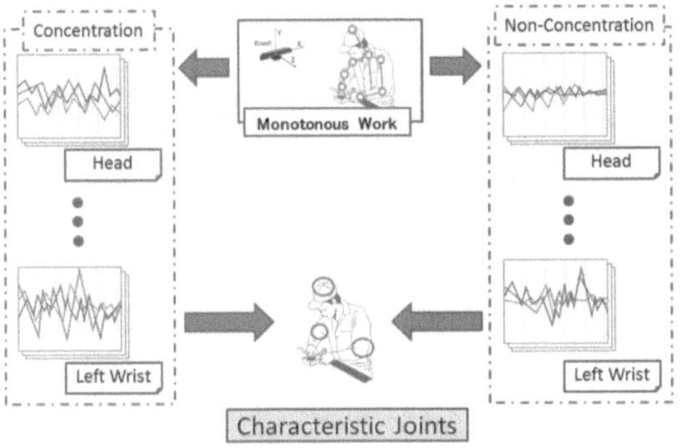

Figure 1. Method overview.

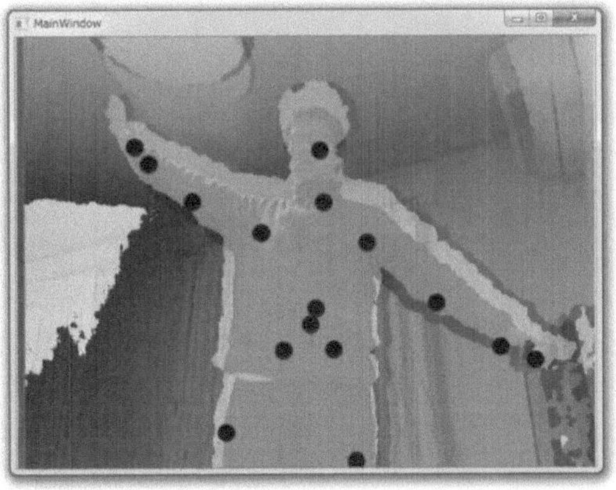

Figure 2. Image example of the Kinect sensor.

of image depths. Fig. 2 shows the image example from the Kinect sensor.

Each black point corresponds to a joint of the person. The Kinect sensor can measure joint position of three directions of twenty black points of the whole body at the same time. Three directions includes the horizontal direction of the image corresponding to the x-axis, the vertical direction to the y-axis, and the depth direction to the z-axis, respectively. The Kinect sensor measures the position of each joint, when a worker repeats monotonous tasks in front of the Kinect sensor. In a line production, every worker takes specific actions for a target lot moving on a conveyer belt. One task is regarded as the actions for a single lot. In the paper, a movement record is a series of time-stamped position of each joint during a task. The number of movement records is in proportion to the number of joints measured during a task. A movement record of each joint is classified into 2 states: a state where a worker is concentration, and the other where a worker is out of concentration.

3.3 *Characteristic sampling*

This method determines characteristic joints, focusing on a fact that repeated monotonous tasks for an excessive long time in a line production would take concentration from workers. In a line production, a worker repeats a same single task monotonously. Repetition of period of regular movement, as the worker concentrates, is because movements of each task become similar as in Fig. 3.

Meanwhile, when the concentration of a worker declines, irregular task periods appear because of actions irrelevant to the task and disorder in the pace of proper actions in the tasks. It indicates that examination of the periodicity in movement records detects the decline of concentration. This method extracts the periods of each joint position in movement records, using the Fourier transformation.

Movement records acquired from tasks with/without concentration is converted into spectrums with the Fourier-transformation. The Fourier transformation takes the position transition of a joint in a specific direction in a movement record as its input, put its spectrum as the output. In this study, we refer the output data as a task spectrum. As the result of the Fourier transformation, we get task spectrums as many as joints of a worker in a task. The task spectrums are given to machine learning as its input. Fig. 4 shows its input data. In the machine learning, a software module is trained so that it should discriminate task periods under concentration from ones out of concentration. The software module is a classifier of task spectrums. Given a new task spectrum, the classifier systematically tells whether it is a task period under concentration or one out of concentration. Characteristic joints should be joints which discriminates the former from the latter better than others. Since characteristic joints vary with workers, we should select the best joints from many ones, according to each workers.

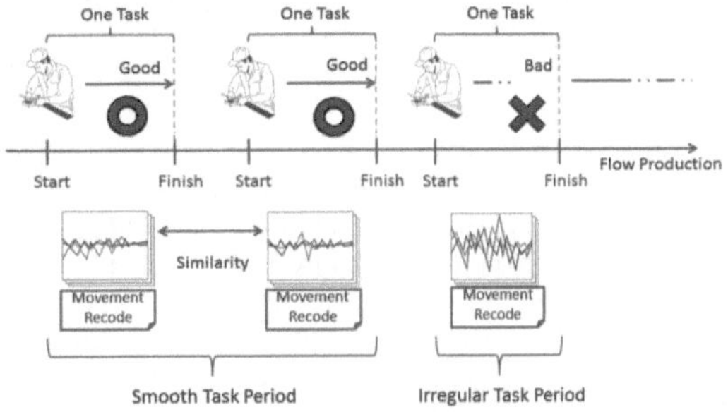

Figure 3. Period of regular movement.

This method uses Random Forest [5] among various machine learning algorithms, because of its evaluation function of each explanatory variable. Since Random Forest scores explanatory variables, it weights the degree of importance of explanatory variables. Any other machine learning algorithm does not have this function.

In this method, explanatory variables used in Random Forest algorithm are amplitude values corresponding to various frequency in the three directions in a task spectrum, as Fig. 4 shows.

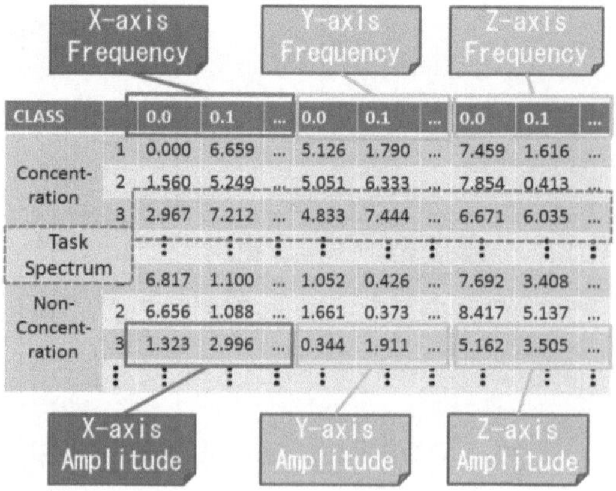

Figure 4. Input data.

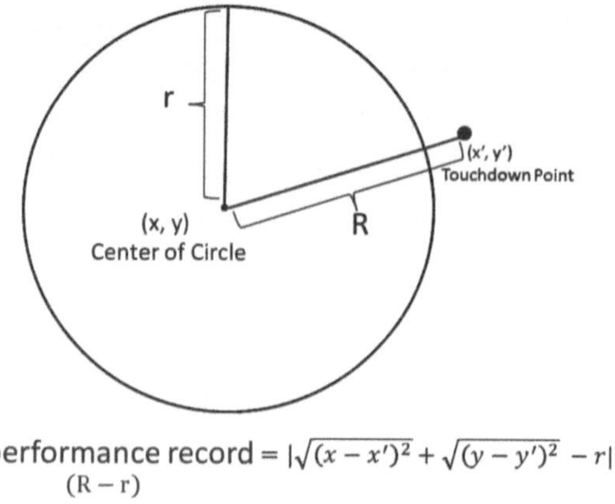

Figure 5. The overview of the performance record.

For each joint, the trained classifier judges whether task spectrums from it belong to ones under concentration or ones out of concentration. Since correct answers are known for all task spectrums, the accuracy rate for the judgment can be calculated for every joint. When the accuracy rate is high for a specific joint, the joint identifies task spectrums of concentration and non-concentration more correctly than other joints. Characteristic joints are joints whose accuracy rates is high. Generally, classification problems of machine learning are required to consider generalization ability. If generalization ability is high, both of recall and precision are high. This method calculates the f-values for each joint with the recall ratio and the precision ratio to determine characteristic joints. In this method, joints having high f-values are adopted as characteristic joints.

4 EXPERIMENTATION

4.1 *Experiment purpose*

An experiment has been conducted to verify whether the proposed method can determine characteristic joints. The purpose of this study is to automatically determine characteristic joints of each worker during monotonous tasks. The F-values is calculated for a classifier trained with task records acquired from many joints of a worker. It is verified the proposed method should find joints effective to discriminate task records under concentration from ones out of concentration.

4.2 *Experiment environment*

In the experiment, testees are seventeen people in total. All of them are twenties; fifteen men and two women.

In the experiment, testees perform the task similar to actual monotonous tasks. The task is to trace a circle displayed on a tablet PC with a stylus within 10 second. A task starts from the touch of the stylus to the screen of the tablet PC. Tasks are regarded to finish with failures, when testees protrude from the circle more than a specific threshold, or do not trace the circle within 10 second. In this experiment, the distance from the center of the circle to the pen top on the screen of the tablet personal computer is measured to represent how testees protrude. The sum of the distance during tracing of one circle is calculated as a performance record. If the disturbance in one task is lower than a specific threshold, the performance record of the task is regarded as a good one, otherwise a poor one. Fig. 5 shows the overview of the performance record.

A Kinect sensor measures the movement of each joint of a testee during tracing of one circle. This experimentation targets 10 joints of the upper half of body, because all testees sit at the desk to conduct the tasks. Since Kinect is a camera-based sensor, overlapping joints produces a blind spot. Therefore, in this experiment, the Kinect sensor is placed at the point at 140 cm in front of testees, higher than the table by 41.5 cm. The head of the Kinect sensor declines by 10 degree. However, the placement causes the distortion of 10 degrees of the measured values from the real one in the y-axis and the z-axis of the Kinect sensor. The corrected value y and z are derived from the measured value y and z with

$$z = z'\cos 10 - y'\sin 10 \qquad (1)$$

$$y = z'\sin 10 + y'\cos 10 \qquad (2)$$

4.3 *Experiment process*

We describe the procedure for testees to take in the experiment environment of explained in section 4.2. Testees sit at a desk where a tablet PC is placed on the table. They adjust the height of the chair so that the Kinect sensor faces each of them from the same direction. When the measurement starts, testees should be accustomed with the monotonous task to tracing a circle sufficiently. Testees repeat 200 times of the monotonous tasks to be accustomed before the measurement starts.

4.4 *Experiment result*

The movements of ten joints of each testee are measured in this experiment. The Random Forest algorithm is applied toward each of ten joint movement to get a classifier. We have analyzed 60 task spectrums, which consist of the best 30 ones and the poorest 30 ones, from the view point of their performance records. 6-fold cross-validation is used for 60 task spectrums of each joint of testees, that is, 50 task spectrums are used for the training of a classifier, while the classifier is tested with the remaining 10 task spectrums in each time of the validation. The 10 task spectrums for the test consist of 5 good ones and 5 poor ones in its performance.

Joints with high F-values are regarded as characteristic joints. A high threshold to determine characteristic joints filters out a few characteristic joints, while a low threshold lists up many. The bias of the position of many characteristic joints in each testee is expected to contribute to finding the similarity of them. This study uses the low threshold 0.6 of F-value. Listing up many characteristic joints with the low threshold, their positions are examined for each testee. Table 1 shows the F-values in each test. In the table, F-values over .6 are colored. The experiment result shows characteristic joints vary with each testee. The result has revealed the effectiveness of the proposed method which finds characteristic joints depending on each worker.

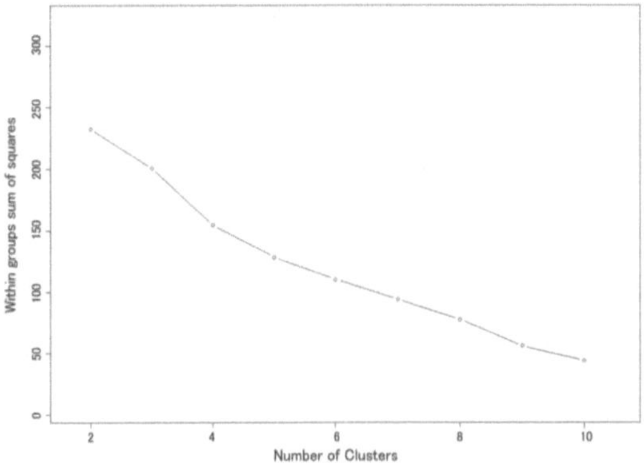

Figure 6. The overview of the performance record.

5 DISCUSSION

5.1 *Clustering*

In the experiment, it is necessary to accumulate movement records, make each worker repeat monotonous tasks in order to determine characteristic joints. However, much effort is imposed on each worker for the repetition.

It cost too much to determine characteristic joints of each worker. A simple way is desired to find joints on which accelerometers are attached for each worker. From manual observation, we have known there are several types for tracing a circle; some testees move only their arm, while others move their arm in conjunction with their whole body. For instance, body trunks of some testees are stable under concentration, while unstable out of concentration. Those testees have characteristics joints all over the body. It can be assumed that some of workers have the similarity in the position of characteristic joints. Suppose workers are classified with the similarity in their body movement at concentration and non-concentration in advance. The judgment of a group a specific worker belongs to would determine characteristic joints the worker should wear accelerometers on. We consider to classify workers from the view point of the similarity. Focusing on the F-values in table 1, we have classified the testess with the k-means algorithm [6]. In the

Table 1. Margin settings for A4 size paper and letter size paper.

	Cluster Number	1	2			3					4							
Joints	Class	H	G	L	N	A	F	J	M	O	B	C	D	E	I	K	P	Q
HandLeft	Concentration	.687	.542	.516	.759	.517	.492	.475	.596	.413	.525	.509	.610	.508	.615	.596	.656	.509
	Non-concentration	.604	.557	.483	.774	.548	.475	.492	.635	.351	.508	.585	.623	.456	.545	.635	.644	.585
WristLeft	Concentration	.625	.702	.655	.780	.533	.508	.483	.571	.467	.500	.600	.536	.618	.557	.557	.625	.431
	Non-concentration	.571	.730	.677	.787	.533	.456	.516	.526	.467	.429	.600	.594	.500	.542	.542	.571	.580
ElbowLeft	Concentration	.635	.556	.754	.519	.406	.429	.516	.500	.459	.627	.491	.491	.476	.523	.688	.500	.351
	Non-concentration	.596	.636	.746	.606	.321	.500	.483	.563	.441	.528	.540	.540	.421	.436	.643	.563	.413
ShoulderLeft	Concentration	.667	.567	.646	.552	.590	.429	.533	.462	.525	.473	.492	.381	.459	.531	.500	.576	.667
	Non-concentration	.667	.567	.582	.581	.576	.500	.533	.588	.508	.554	.475	.316	.441	.464	.500	.590	.632
ShoulderCenter	Concentration	.677	.526	.452	.567	.508	.370	.613	.426	.571	.478	.475	.429	.700	.492	.475	.594	.613
	Non-concentration	.655	.571	.414	.567	.525	.485	.586	.630	.625	.340	.492	.500	.700	.400	.492	.536	.586
Head	Concentration	.627	.433	.540	.509	.545	.393	.585	.440	.545	.441	.571	.421	.448	.459	.426	.433	.613
	Non-concentration	.528	.433	.491	.585	.615	.469	.509	.600	.615	.459	.526	.476	.484	.441	.407	.433	.586
HandRight	Concentration	.636	.698	.656	.786	.407	.467	.552	.540	.508	.567	.491	.333	.567	.431	.545	.493	.491
	Non-concentration	.556	.667	.607	.813	.426	.467	.581	.491	.525	.567	.540	.455	.453	.327	.444	.314	.540
WirstRight	Concentration	.738	.600	.646	.793	.381	.179	.586	.526	.557	.551	.436	.154	.554	.531	.533	.533	.340
	Non-concentration	.691	.600	.582	.806	.316	.281	.613	.571	.542	.392	.523	.353	.473	.464	.533	.533	.478
ElbowRight	Concentration	.781	.571	.633	.586	.606	.621	.667	.582	.623	.563	.509	.593	.500	.563	.457	.596	.484
	Non-concentration	.750	.526	.633	.613	.519	.645	.632	.646	.610	.500	.585	.667	.500	.500	.240	.635	.448
ShoulderRight	Concentration	.667	.536	.557	.623	.554	.373	.656	.596	.581	.379	.533	.367	.475	.400	.357	.576	.373
	Non-concentration	.698	.594	.542	.610	.473	.393	.644	.635	.552	.419	.533	.367	.492	.492	.438	.590	.393

classification, characteristic joints whose F-value is more than .6 are regarded as strong ones. Fig. 6 depicts the graph indicating the relationship between the number of clusters and the sum of squared distance within groups.

When the sum of squared distances within a group gets smaller, elements into the group are very similar with each other. The more groups get, the smaller the sum of squared distance gets. Meanwhile, the more groups are prepared, the harder it is to classify workers into the groups. To take an appropriate balance, we should determine a suitable number of groups. This discussion adopts 4 clusters just after the sum of squared distances goes down rapidly. Table 1 shows the classification result with the F-values, where joints with the F-value over .6 are colored.

Table 1 shows the 4 clusters below. Workers belonging to Cluster 1 have characteristics to over the whole bodies. In Cluster 2, characteristics are found in joints of both arms. Cluster 3 has characteristics in joints of their right arms. Cluster 4 workers show has characteristics in their left arms.

The analysis has narrowed the groups based on the similarity in characteristic joints up to no more than 4. It is possible to classify workers into the groups with tasks far less than those in the experiment to precisely identify their characteristic joints. A method to classify workers using data collected with Kinect would eliminate the necessity of the labor-intensive experiments for each worker.

6 CONCLUSION

Every worker is assumed to have his own characteristic joints showing different movement in case he engages in tasks under concentration and in case out of it. Accelerometers attached to these joints contribute to preventing human errors which brings serious injuries and tremendous loss caused by careless mistakes. The method this paper proposes determines characteristic joints depending on each worker and each task with a Kinect sensor in monotonous tasks. In this method, the Kinect sensor gets movements of workers during tasks they engage in both under and out of concentration. The movement records acquired during tasks are converted into task spectrums with the Fourier-transformation. A classifier trained by Random Forest using many task spectrums of each joint distinguishes movement records under concentration from ones out of concentration, to determine characteristic joints of a specific worker. To confirm whether we can determine characteristic joints with this method, we have conducted an experiment, where testees repeat tasks to trace a circle on the tablet PC in front of the Kinect sensor. Judging whether movement records are ones under concentration or ones out of it with the classifier, the F-values of each joint of testees are calculated. As a result, almost all the testees have characteristic joints. We have found all of them with the method. We plan to apply this method to other tasks to show the effectiveness in general use.

REFERENCES

[1] K. Shinohara, K. Morimoto, T. Kubota; "The ef fect of "Finger-pointing and Call" on orientation of visual attention" (in Japanese). The Japanese journal of ergonomics, 45(1), 54–57 (2009)

[2] Y. Tontani, Y. Kajiwara, F. Harada, H. Simakawa; "Judging Working Rhythm from Body Movement to Prevent Human Errors". Proc. of the 12th International Conference on Telecommunications and Informatics, pp. 127–132 (Sep. 16–19, 2013, Baltimore, US)

[3] K. Kousaka, K. Imaki, M. Shibata, H. Haga, K. Shigeo; "Classification of Children's Group Activity from Acceleration Data by using Wavelet Transformation" (in Japanese). UBI, 2009-UBI-22(13), 1–8 (2009)

[4] S. Kaneda, K. Kosaka, K. Imaki, T. Mitsumoto, K. Shintani, A. Kono; "A Concentration Analysis Approach for Storytelling using Stereo Cameras and Accelerometers" (in Japanese). IPSJ SIGCE, 2010-CE-104(2), 1–8 (2010)

[5] Breiman, Leo; "Random Forests". Machine Learning, 45(1), 5–32 (2001)

[6] J. McQueen; "Some methods for classification and analysis of multivariate bservations". In Proceedings of the Fifth Berkeley Symposium on athematical Statistics and Probability, pp. 281–297 (1967)

Industrial Engineering and Management Science – Lee (Ed)
© 2015 Taylor & Francis Group, London, ISBN 978-1-138-02647-6

A system for recognizing IADL using brightness-distribution sensors

S. Shimayoshi
Graduate School of Science and Engineering, Ritsumeikan University, Shiga, Japan

Yusuke Kajiwara & Hiromitsu Shimakawa
College of Information Science and Engineering, Ritsumeikan University, Shiga, Japan

ABSTRACT: In Japan, there are many elderly persons who live alone and die because of natural causes at home without being known to anyone. To prevent dying alone, nursing staffs need to comprehend mental and physical health of the elderly persons using remote life monitoring system. This paper presents a remote life monitoring system to comprehend mental and physical health of the elderly persons, recognizing their Instrumental Activity of Daily Living (IADL) using Brightness-Distribution Sensors (BDS). As an experiment result, the proposed system is able to identify a person learning IADL of some persons in advance. However, the proposed method has to learn the way of cleaning in each person before mental and physical weakness, because the way of cleaning is different in each person.

1 INTRODUCTION

In Japan, it is a serious social problem that elderly persons live alone and die because of natural causes at home unattended. To prevent dying alone, nursing staffs need to comprehend mental and physical health of the elderly persons who are getting weak using life monitoring system.

This paper presents a life monitoring system to comprehend mental and physical health of the elderly persons recognizing Instrumental Activity of Daily Living (IADL), such as cleaning, laundry, and kitchen work using a Brightness-Distribution Sensor (BDS) (Nakashima, S. et al. 2010). Since the elderly persons get weak physically and mentally, they take less and less IADLs in their daily life. The detection of the frequency of IADL is expected to work as criteria to detect their decline. Especially, cleaning is a typical activity to judge their activeness, because they can live without it, while it is very annoying. Remote monitoring of cleaning activity contributes to detection of the decline. However, the way of cleaning strongly depends on characteristics of each elderly person. The paper examines ways of cleaning for various persons to know personal cleaning characteristics. Personal characteristics enable the life monitoring system to learn a way of cleaning in each elderly person so that it can detect mental and physical weakness individually.

2 PRECEDENT STUDY

Precedent studies recognize Activity of Daily Living (ADL) using image sensor (SEKI, H. et al. 2002), optical sensors (Saito, M. et al. 2008), infrared sensors (Kajiwara, Y. et al. 2012), or pressure distribution sensor (Tajima, T. et al. 2011). Image sensor invades privacy of elderly persons. On the other hands, optical sensors, infrared sensors and pressure distribution sensor protect privacy of elderly persons than image sensor. Optical sensors weaken environmental changes, infrared sensors need us to arrange a multitude of sensors in the house, and pressure distribution sensor is expensive. For these reasons, the introduction of above sensors is difficult for elderly persons.

On the other hand, BDS protects privacy of elderly persons. It is able to measure wide scope and yet inexpensive.

3 PROPOSED SYSTEM

This study proposes a life monitoring system to recognize IADL of elderly persons using BDS. Figure 1 shows life monitoring system acquires brightness distribution value from $BDS^{(1)}$ and $BDS^{(2)}$. BDS is arranged in the house of an elderly person. BDS recognizes IADL using machine learning algorithms. The superscript character is the specific number of each other.

3.1 *Acquiring brightness-distribution value*

Figure 2 shows BDS acquires brightness distribution value $v^{*}_{c,x}$, $(c = 1, 2, \ldots, C, x = 1, 2, \ldots, X)$ which integrates pixel of y-axis to film objects using a rod lens. From the brightness distribution value can't be restored a pixel in the direction of the y-axis and original image. Therefore, BDS protects the privacy

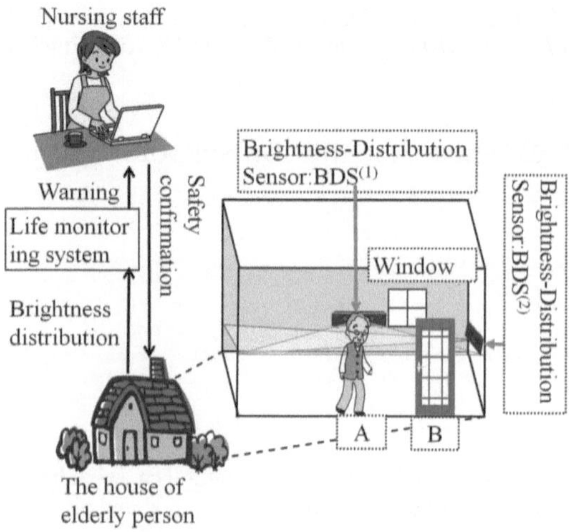

Figure 1. Life morning system using brightness-distribution sensors.

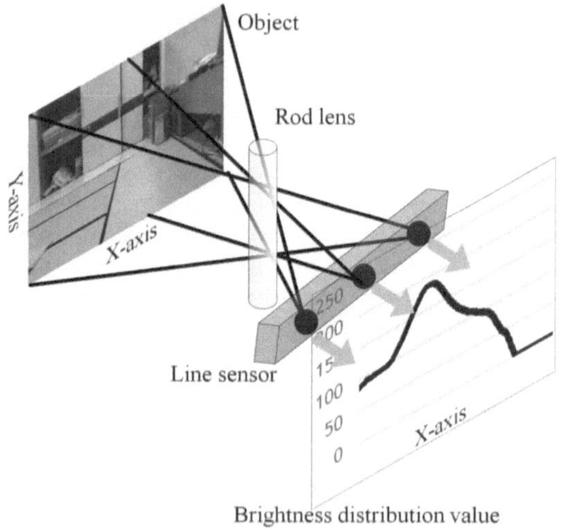

Figure 2. Brightness distribution sensor and its structure.

of the elderly person, but image sensor does. Two brightness-distribution sensors are arranged orthogonal to each other to measure activity of daily living of elderly person in living room.

Proposed system recognizes IADL from X (=640) brightness distribution value between C (=600) frames. Frame rate is 30 fps.

3.2 *Back ground difference method*

Figure 3 shows an outline of background difference method. This method calculates background difference value $d^*_{c,x}$ from brightness distribution value $v^*_{BG,x}$ ($x = 1, 2, \ldots, 640$) of image without human figure. $v^*_{c,x}$ is calculated as below:

$$d^*_{c,x} = | v^*_{\mathrm{BG,x}} - v^*_{c,x} | \tag{1}$$

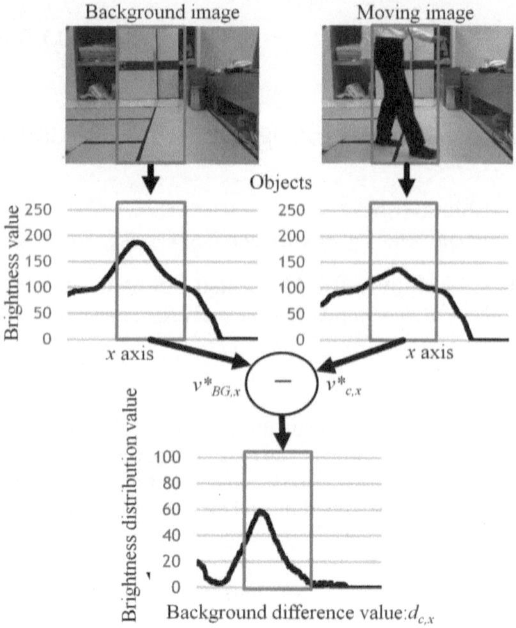

Figure 3. Background difference methods.

Background difference value $d^*_{c,x}$ changes when a person acts.

3.3 *Bag of features*

Feature vector is calculated to input data into Bag of Features (BoF) (Csurka, G. et al. 2004). The input vector $I^*_{c,x}$ has $d^*_{c,x}$, space difference $dp^*_{c,x}$, and time difference elements $dc^*_{c,x}$. $dp^*_{c,x}$ and $dc^*_{c,x}$ calculate as below

$$dp^*_{c,x} = | d^*_{\mathrm{c,x}} - d^*_{c,x-1} | \tag{2}$$

$$dc^*_{c,x} = | d^*_{\mathrm{c,x}} - d^*_{c-1,x} | \tag{3}$$

BoF calculates K (=25) center clusters for input $I^*_{c,x} = [d^*_{c,x} \ dp^*_{c,x} \ dc^*_{c,x}]^T$ ($c = 1, 2, \ldots, 60$, $x = 1, 2, \ldots, 640$) into k-means++ methods (Arthur, D. et al. 2007) . Euclidean distance between K center clusters and $I^*_{c,x}$ is calculated, BoF creates a K-dimensional histogram to vote for the nearest center cluster on the basis of the Euclidean distance between K center clusters and $I^*_{c,x}$. Figure 4 shows K-dimensional histogram $hist^*_i$ ($i = 1, 2, \ldots, 6$) is calculated from the brightness distribution values which are divided into areas in the space. Feature vector has $6 \times 25 \times 2$ elements for two brightness distribution sensors.

3.4 *Identification by random forest*

Random forest (Breiman, L. 2001) recognizes IADL to input feature vector. Random Forest creates T decision trees, and outputs the majority result in T

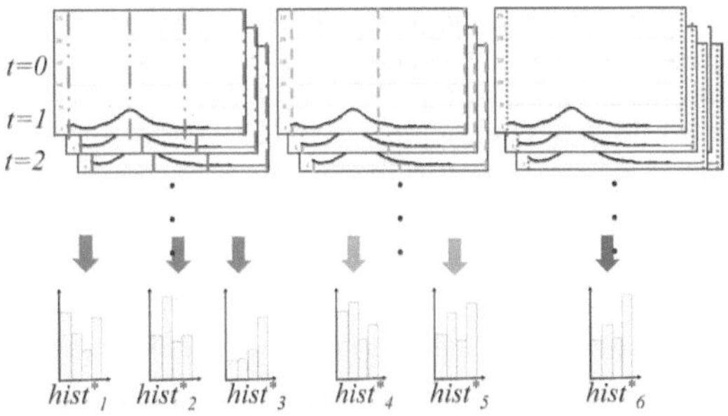

Figure 4. K-dimensional histogram is calculated in each area in the space.

decision trees. Decision trees are created using CART (Breiman, L. et al. 1984). CART recognizes IADL on the basis of explanatory variables in which Gini's coefficient become maximum. Variable importance is the value of summing up Gini's coefficient in each explanatory variable.

4 EVALUATION EXPERIMENT

4.1 *Experiment method*

Identification accuracy is calculated. In addition, way of cleaning is analyzed form the differences among individuals by discrimination analysis. Identify actions such as walking (WK), hanging the laundry (HL), taking the laundry (TL), cleaning (CN). There are 6 subjects, 5 subjects are male, and one subject is female. The subjects act above IADL at 20 times. Each IADL acts as below.

(a) Walking
The subject enters room from place A in Figure 1.
The subject stops near a window.
The subject leaves room from place B in Figure 1.

(b) Hanging the laundry
The subject having a laundry basket enters room from place A in Figure 1.
The subject stops near a window.
The subject hangs the laundry in a laundry basket at the window.
The subject having a laundry basket leaves room from place B in Figure 1.

(c) Taking in the laundry
The subject having a laundry basket enters room from place A in Figure 1.
The subject stops near a window.
The subject takes in the laundry at the window.
The subject having a laundry basket leaves room from place B in Figure 1.

(d) Cleaning
The subject having a vacuum cleaner enters room from place A in Figure 1.
The subject cleans room with the vacuum cleaner.
The subject having a vacuum cleaner leaves room from place B in Figure 1.

Table 1. The identification result of IADL.

	WK	HL	TL	CN	R	P	F
WK	101	11	3	0	0.84	0.88	0.86
HL	17	55	33	0	0.46	0.52	0.49
TL	2	54	84	0	0.70	0.60	0.65
CN	0	0	0	120	1.00	1.00	1.00

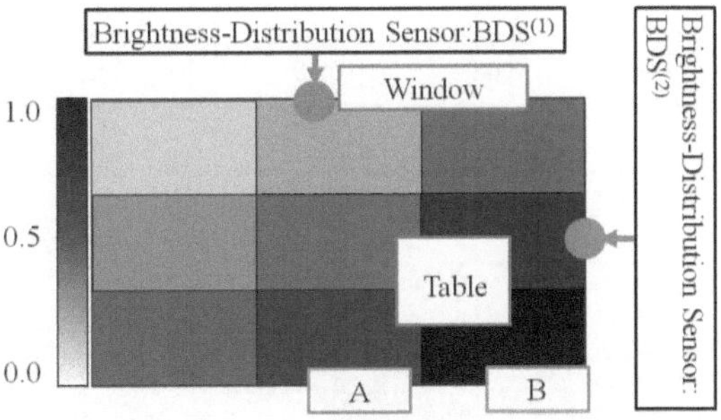

Figure 5. Variable importance of each explanatory variable when proposed system identifies IADL.

The subjects train each IADL before beginning this experiment in order to get used to each action. The proposed system is evaluated with 6-fold cross validation, regards test set as a feature vector of a person, and regards training set as a feature vector of other persons. A way of cleaning on differences among individuals is analyzed 20-fold cross validation. Regards test set as one of cleaning in each subject, and regards training set as other cleaning in each subject. If the way of cleaning is different in each person, the persons can be specified by the way of cleaning. Evaluation value is F-measure F:

$$F = \frac{2 \times R \times P}{(R + P)} \tag{4}$$

where R is recall, P is precision.

4.2 *Experiment result and discussion*

Table 1 shows identification accuracy of each IADL. CN is able to completely identified other actions, this result shows that proposed system is able to identify a person learning IADL of some persons in advance. On the other hand, HL and TL are different actions at the window, which are misidentified with each other. This result shows that fine action as HL and TL can't be measured using BDS due to spatial resolution. Figure 5 shows variable importance of each explanatory variable when proposed system identifies IADL. Highly variable importance is acquired near place B in Figure 5. Because many persons only leave or clean the room near an exit, and rarely do other actions.

Table 2 shows the identification accuracy of person from way of cleaning. The average of f-measure is 0.9, the way of cleaning is different from each person.

Table 2. The identification accuracy of person from way of cleaning.

Subject	Recall	Precision	F-measure
A	0.90	0.95	0.92
B	0.90	0.75	0.82
C	0.75	0.88	0.81
D	0.95	0.90	0.93
E	0.90	1.00	0.95
F	1.00	0.95	0.98
AVG	0.90	0.91	0.90
SD	0.08	0.09	0.07

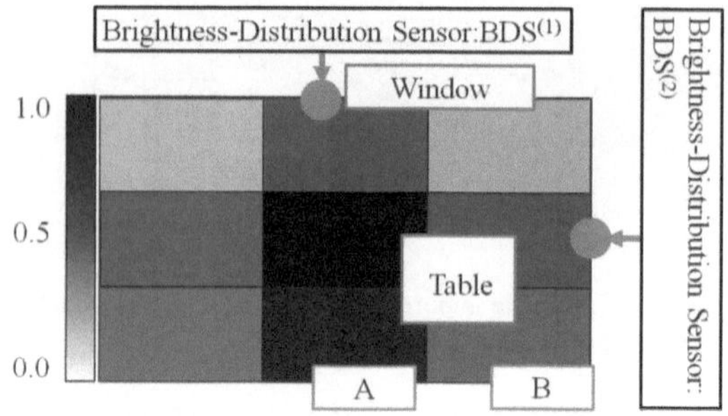

Figure 6. Variable importance of each explanatory variable when proposed system identifies persons.

Therefore, the life monitoring system needs to learn way of cleaning in each elderly person before mental and physical weakness. Figure 6 shows variable importance of each explanatory variable when proposed system identifies persons. High variable importance is acquired near place A in Figure 6. Because place A in Figure 6 area is larger than place B in Figure 6 area.

5 CONCLUSION AND FUTURE WORK

The proposed system is able to identify a person learning IADL of some persons in advance. However, the proposed system learns the way of cleaning in each person before mental and physical weakness due to the way of cleaning difference in each person. As future work, the way of cleaning is analyzed with each person, we creates an evaluation function regarding mental and physical weakness.

REFERENCES

Arthur, D. & Vassilvitskii, S. 2007. k-means++: the advantages of careful seeding, *Proceedings of the eighteenth annual ACM-SIAM symposium on Discrete algorithms*: 1027–1035.

Breiman, L. 2001. Random Forests, *Machine Learning*, 45(1): 5–32.

Breiman, L., Friedman, J., Olshen, R. & Stone, C. 1984. Classication and Regression Trees, *Wadsworth*.

Csurka, G., Dance, C.R., Fan, L. & Bray, C. 2004. Visual Categorization with Bags of Keypoints, *European Conference on Computer Vision*: 1–22.

Kajiwara, Y., Kobayashi, Y., Nakamura, M., Hayashi, K. & Kimura, H. 2012. Development of a fall detection system in bathroom using photodetectors. *IEEJ-E* 132(11): 387–396.

Nakashima, S., Kitazono, Y., Zhanga, L. & Serikawa, S. 2010. Development of privacy-preserving sensor for person detection, *Procedia Social and Behavioral Sciences* 2(1): 213–217.

Rimminen, H., Lindstrom, J., Linnavuo, M. & Sepponen R. 2010. Detection of Falls Among the Elderly by a Floor Sensor Using the Electric Near Field, *IEEE Transactions on Information Technology in Biomedicine* 145(6): 1475–1476.

Saito, M., Kitazono, Y., Seiichi, S. 2008. Development of Sensing System for the Estimation of Human's State Using Infrared Sensors Arranged at Grid, *IEEJ Trans. SM* 128(1): 24–25.

Seki, H. & Hori, Y. 2002. Detection of Abnormal Action Using Image Sequence for Monitoring System of Aged People. *IEEJ-D* 122(2): 182–188.

Tajima, T., Abe, T. & Kimura, H. 2011. Development of Fall Detection System Using Ultrasound Sensors, *IEEJ-E* 131(1): 45–52.

Industrial Engineering and Management Science – Lee (Ed)
© 2015 Taylor & Francis Group, London, ISBN 978-1-138-02647-6

Estimating watering amount with soil moisture sensors to identify farming works

S. Nakanishi & K. Yasui
Graduate School of Information Science and Engineering, Ritsumeikan University, Shiga, Japan

Y. Kajiwara & H. Shimakawa
College of Information Science and Engineering, Ritsumeikan University, Shiga, Japan

ABSTRACT: Recently, many people want to start farming. Novice farmers learn techniques of experienced ones from their works. Many techniques in farming are acquired on experiences and intuition, thus it is difficult for novice farmers to get the techniques. To solve this problem, objective agricultural techniques should be extracted from quantitative information instead of subjective one. The quantitative information includes not only data acquired from farming fields with sensors but also records of farming works. However, periodical interviews or questionnaires for the records on farming works interfere with daily activities of farmers. Farming works should automatically be recorded to reduce burdens of farmers.

This research focuses on watering farmers' daily conduct, which has great effects on crops. This paper proposes a method to estimate the amount of watering with soil moisture sensors, distinguishing the increase of soil moisture from other reasons. Watering causes rapid increase of soil moisture. The method records transitions of soil moisture with a sensor measuring the volume water content, which indicates the amount of water in a unit volume of soil. A regression curve is calculated from the transitions. Actual measurements have errors from estimated values through the regression curves. If the rate of error occurrence during the transition stays within a specific range, we can make a decision watering is conducted with a corresponding amount of water.

We have conducted an experiment to estimate the amount of watering from transitions of the volume water content, changing the water volume by 3 levels. Optimal thresholds for the error range and the rate estimate the watering amount with the accuracy of about 72%.

1 INTRODUCTION

Recently, many people want to start farming. In Japan, the number of consultations to start farming has doubled from 2007 to 2009. However, the number of new farmers has not increased, which account in some degree for Japanese farming decline. It is a vital issue to increase new farmers and train novice farmers in Japan. Novice farmers should learn techniques to start farming. It is difficult for many novice farmers to learn the farming techniques, which in a big obstacle for them to start farming.

Current farming techniques for crop cultivation are generally subjective, because they rely on experiences and intuition. For example, farmers judge a state of crops by observing leaf color. If its color is bad, farmers add fertilizer on the field. The criteria to make judgment are subjective, hence the methods are not available to all farmers. Even experienced farmers miss the judgment so that their crops get worse in quality and quantity. It is large economic risk for farmers, and prevents people from starting farming. In addition, it is hard for experienced farmers to teach farming techniques to novice farmers, because farming techniques relays on subjective information. It is essential to establish standards based on objective information on when and what farm work they take. It is necessary to facilitate the learning techniques to get more novice farmers.

In this paper, we focus on watering which is a basic farm work. A method is proposed to estimate watering amount with soil moisture sensors. We have experience to determine thresholds through the prediction based on the regressive analysis. In this experiment, a soil moisture sensor was used for data acquisition from soil in planters. Optimal thresholds estimate the watering amount with the accuracy of about 72%.

2 ACQUISITION OF WORK INFORMATION FROM FARMING FIELD

2.1 *Formalization of farming techniques*

To teach the novice farmers farming techniques, it is not preferable to judge the states of farming field and farm work timing based on experiences and intuition. It should be changed to the objective judgment based on the quantitative information. This is called the formalization of farming techniques. For examples, farmers use fertilizer if leaf colors are bad, and irrigate if field soil is dry.

We focus on these farming techniques. The judgments on leaf colors and soil states are subjective ones of farmers. These subjective techniques are based on experiences and intuition of farmers, which leads to mistakes in preventing problem of crops appropriately. Since they rely on experience and intuition, it is difficult for novice farmers to get the farming techniques. The change of subjective techniques into objective one requires to obtain the quantitative information from fields: the crop states such as size and color, temperature and soil moisture. The farmers would take accurate farming work with a manual. For example, they should take watering in particular soil states judged from relationships of states of crops with that of a farming field. In addition, the novice farmers can judge without experiences and intuition because they can use the manual which relies on quantitative information.

2.2 *Acquisition of farming field states*

To extract formal farming techniques, each states of farming field and crops should be obtained quantitatively in each farm works. Farming techniques of experienced farmers are tried to be formalized analyzing the data before and after a farm work [1]. The quantitative farming techniques are obtained extracting information which experienced farmers rely on to judge timing of farm work using relation of states of farming field with that of crops. Farming field sensors and cameras obtain the states of farming field automatically. However, the farmers must record farm works, such as information like watering timing and amount with their own hands. It makes farmers fatigued.

2.3 *Exiting works*

Some works propose methods to reduce the burden of farmers when recording the farm work information. Farmers record the farm work information with computers and smart phones [2]. However, the farmers are burdened with data input because they should record the information into computers in timing of performing the farm work. To reduce burden of farmers, farm work information must be recorded automatically without manual tasks of farmers. Another research uses RFID to record the farm work information automatically [3]. This method obtain the time and kind of farm work to read RFID on farming implements and facilities with RFID reader farmers wear on their wrists. However, farmers must wear the RFID reader during farm working to obtain the information. It prevents farmers from performing farm work. It is necessary to record and obtain farm work information without burdens of farmers.

3 ESTIMATING WATERING AMOUNT WITH SOIL MOISTURE TRANSITIONS

3.1 *Soil moisture transitions by watering*

The farm works include watering, fertilization, and ventilation if farmers culture crops in a greenhouse. They perform watering every day in the farm work. In addition, watering greatly affects crops. The watering imposes big burden on farmers to record the information because the watering is performed frequently than the fertilization and ventilation. If automatic recording of watering is possible, it can reduce the burden of farmers when they record farm work information. In this research, we focus on watering that is the most important in farm work.

Our approach focuses on change of soil moisture by watering to obtain watering information, because watering affects transitions of soil moisture. However, rainfall and rising of ground water affect soil moisture as well. Watering uses a lot of water for a short time. We expect that soil moisture increases rapidly on watering. In contrast, the soil moisture changes slowly in rainfalls. The rise of ground water does not affect increase the soil moisture rapidly, neither. The rapid increase of soil moisture with watering only happens at watering. The transition of the transfer of soil moisture depends on a character of soil. For examples, if soil is sandy, the soil moisture of a deep place increases immediately. If soil is clayey, the soil moisture of a deep place increases slowly. We must consider the difference of soil characteristics in each farm field to estimate the watering amount. In this paper, we propose to estimate watering amount from the transition of soil moisture measured with soil moisture sensors.

3.2 *Factors of soil moisture changing*

Factors such as transpiration and soil wetting are used to model soil moisture changing [4].

- Watering
- Rainfall
- Rise of ground water

Figure 1 shows an example of the soil moisture in a greenhouse. The example shows the rise of ground water affects the farming field. In this experiment, water potential sensors are used as an index of soil moisture. The sensors are set below 30 cm of the surface as Figure 1. The water potential sensors show small values for wet soil and large value for dry one.

As Figure 1, the soil moisture does not increase directly with rainfall because the rainfall does not pour on the soil in a greenhouse. Since, the water potential would decrease after few days from the rainfall, we can know the soil moisture increased. We focus on sensor location in greenhouse as Figure 1. The greenhouse slopes gradually from entrance to the drain. The water pools on paths on both sides of greenhouse. The soil moisture has increased in order of point 2 to point 1, but not increased at point 3. The fact suggests that the water pooled on the path wetted the soil slowly. The

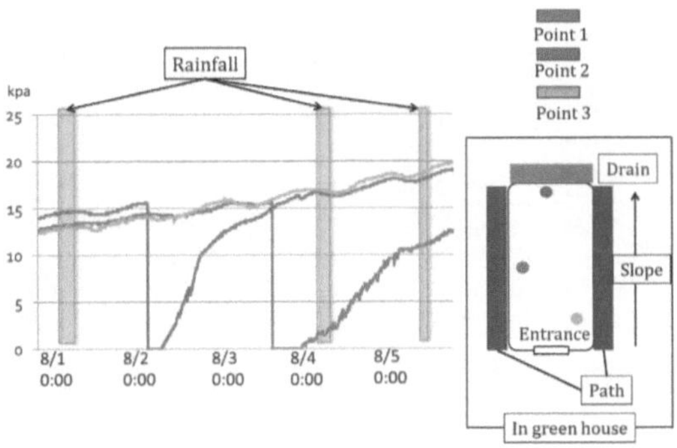

Figure 1. Changing of soil moisture in green house.

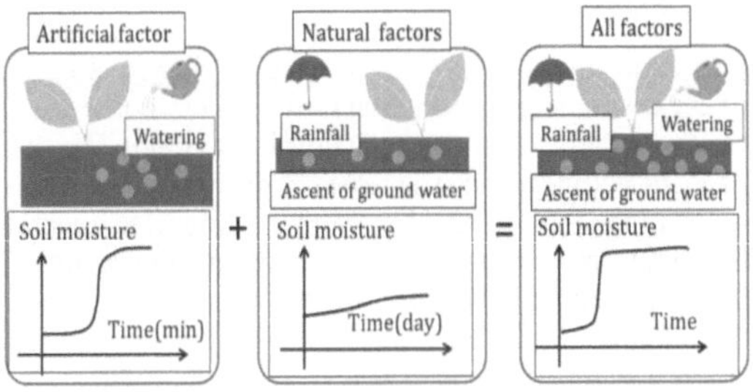

Figure 2. Rise of soil moisture with complex factors.

greenhouse and water in the soil has moved from point 2 to point 1 in accordance with the slope. It is assumed that the rainfall has changed the groundwater, and has increased the soil moisture in greenhouse. As the analysis shows, the soil moisture in farming field increases with many reasons. As figure 2 shows, the soil moisture with many reasons, even though it is measured as soil moisture in the farming field.

It is difficult to distinguish the transition of soil moisture with watering from that of soil moisture from other as all complex factors. However, the transition of the soil moisture seems to be slow without watering. We assume to be able to classify the transition of watering and the transitions without watering as figure 2, and identify the character of the transitions. Comparing transitions witch the characteristics can distinguish the transition of soil moisture in farming field.

In this research,

- We expect the transition of soil moisture for watering with regression analysis using the data when performed watering in an environment which enclosed artificially other factors.
- We predict the transition of soil moisture without watering factor with regression analysis using the data of an environment without watering factor.
- We focus on characteristics of transition for short time and to identify the watering.

It takes a long time to record the transition of soil moisture without watering factor. In this research, we focus on to regression analysis to predict the transition of watering.

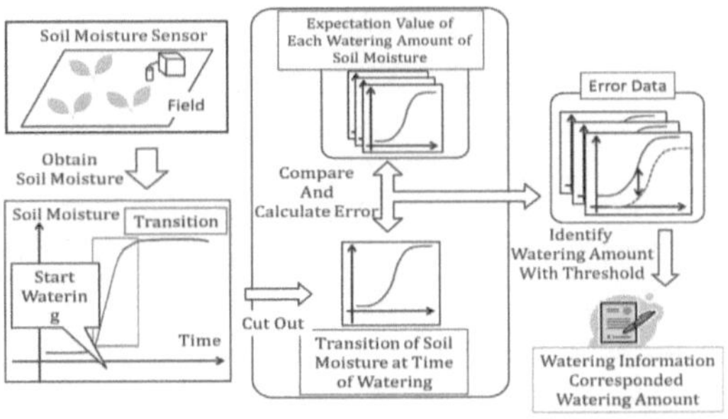

Figure 3. Method of estimating watering amount.

3.3 *Estimating watering amount from transition of soil moisture*

In this method, we focus on the transition of soil moisture at watering for estimating watering amount. Figure 3 shows the outline of the method. The soil moisture increases rapidly when farmers perform the watering on farming field. For measuring the transition of soil moisture at watering, the soil moisture is measured with soil moisture sensor from farming field. The transitions of soil moisture are measured to identify watering amount from the data of soil moisture of field. The data is cut out from the point just before increase of soil moisture to stabilization point for analysis. The data are divided into the training data and the test data. Regression expression is applied for the training data. The regression expression calculates the predicted value of soil moisture at a specified point of time. The error at each point is calculated comparing the test data with the predicted data for the specific amount watering. If the calculated error is within a threshold, we identify to perform the watering with corresponded watering amount. The obtaining farm work information at the time is watering amount. Therefore, the obtaining the watering amount with farm field sensors reduces farmers' burden in recording watering information.

3.4 *Determining the regression curve*

Suppose a cutting out data row is $Y = (y_0, \ldots, y_\omega)$. We assume the change of soil moisture follows a logistical curve. The transition of soil moisture of each watering amount is represented with logistics curve. A logistics curve base of regression curve shows next expression. The a, b and c are parameters of logistics function.

$$f = \frac{1}{1 + e^{\frac{b-x}{c}}} \tag{1}$$

The data row of logistics regression curve is $\hat{Y} = (\hat{y}_{0,\ldots,}\hat{y}_\omega)$. The predicted value is calculated with the logistics curve where parameters are identified with the soil moisture data of each watering amount.

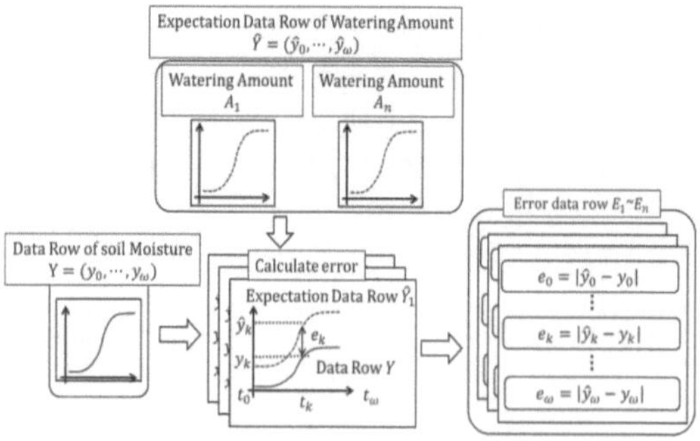

Figure 4. Method of error calculation.

3.5 *Calculation of error and threshold*

Figure 4 shows outline of calculation of the error data row.

Comparing the curve of logistics regression with cut out data, the error is calculated. The data row of the logistics regression curve is $\hat{Y} = (\hat{y}_0, \ldots, \hat{y}_\omega)$, and the test data row for comparison is $Y = (y_0, \ldots, y_\omega)$. The error row $E = (e_0, \ldots, e_\omega)$ of the data row and the data row of logistics regression $\hat{Y}$ is calculated with next expression.

$$e_i = |\hat{y}_i - y_i| \tag{2}$$

$$E = \sum_{i=0}^{\omega} e_i \tag{3}$$

We determine the threshold of the error to identify the watering amount. The threshold W of e_i is determined to identify actual watering amount. Here, d is defined as the number of i such that e_i is within threshold W. The next expression calculates the content percentage r to identify the similarity of the data row Y and the regression data row $\hat{Y}$.

$$r = \frac{d}{\omega} \tag{4}$$

We define the threshold R of the content percentage r for identifying watering amount. If the d and r are within threshold, W and R, we conclude to perform watering with the water amount that corresponding predicted value.

4 EXPERIMENT AND ANALYSIS

4.1 *Used sensor*

In this method, we measure the soil moisture for estimating watering amount. To measure the soil moisture, there is a method that gets and dries the soil. However, it is a burden of farmers. It is difficult to obtain continuously the soil moisture which varies with watering. In our experiment, we use a soil moisture sensor SMEC300 by Spectrum to measure the soil moisture.

Figure 5. Experiment image.

4.2 *Experiment*

Our experiment aims determining the logistics curve of each watering amount that requires our method, and the threshold to identify the watering amount. We prepare for reappearance the states of performed watering to farming field. Sensors are set into the soil to observe the changing of volume water content in percentage. We made the planter of 30 cm square using wood board, and filled the planter with soil of 10 cm height (Figure 5).

Soil of the same farming field is used in this experiment. The change of volume water content of each watering amount was observed at 1 minute intervals for watering with specific water amount onto planter. The watering is performed 5 times with each watering amount in 1 L, 1.5 L and 2 L. Each watering is performed 15 times. The soil is changed to the new soil every time because the initial value of the volume water content change for watering.

4.3 *Analysis*

We have examined whether the watering amount was identified with the data obtained in our experiment. We cut out the transition of volume water content with $\omega = 15$. In analysis of experiment data, the test data is chosen from one transition data of volume water content of five data in three kind of watering amount for true positive distinction. The four others' transition data of volume content are calculated logistics regression curve. The error is calculated to compare the transition data of expectation value that calculated with regression curve and the transition data of volume water content that is one of test data. For true negative distinction, the error row is calculated to compare the expectation value and 10 transition data of volume water content that is the two others kind of watering amount. In this way, the error row is calculated to choose volume water content of each watering amount at five times and total 15 times for true positive distinction and true negative distinction. The threshold W is changed 0% to 10% by 0.1% for the calculated error row, and the threshold R is changed 50% to 100% by 5% for the content rate r. The combination of W and R that are the optimal correct answers rate was calculated to change two thresholds. The percentage of exact distinction of watering amount is defined as the percentage of correct answers C.

$$C = \frac{\begin{matrix} \textit{correct answers number of} \\ \textit{true positive distinction} \end{matrix}}{\textit{correct answers number of true positive}} + \frac{\begin{matrix} \textit{correct answers number of} \\ \textit{true negative distinction} \end{matrix}}{\textit{correct answers number of true negative}} \quad (5)$$

The watering amount is identified the optimal correct answer rate with C = 72% when the threshold met $7.8 \leq W \leq 7.9$ and R = 85%. Therefore, this method can identify the watering amount to measure the volume water content with the soil moisture sensor and analyzed data. This method can obtain the farm work information of watering if the optimal threshold identified the watering amount with accuracy of about 72%.

We can contribute that the formalization of farming techniques to combine the watering information that obtain with this method, the crop states and other farm work information.

5 CONCLUSION

In this paper, we have proposed to estimate the watering amount with soil moisture sensor. In this method, we obtained the transition of soil moisture at watering with soil moisture sensor in farming field. The error at each point is calculated to compare the test data and the expectation data for the specific amount watering. We defined the threshold for the calculated error and identified to perform watering with water amount that corresponded expectation value.

In an experiment, we obtained the water volume content data at the time of watering with changing the watering amount by 3 levels.

The error was calculated with comparing the expectation value of watering amount and the watering data. The combination of thresholds that are the optimal correct answers rate was calculated with changing two thresholds. Optimal thresholds for the error range and the rate estimated the watering amount with the accuracy of about 72%.

REFERENCES

[1] Agriculture, Forestry and Fisheries Research Council http://www.s.affrc.go.jp/docs/e/index.htm

[2] P.K.S.C. Jayasinghe, et al.: "Development of a Fieldwork Reminder System to Help Field Management", World Conference on Agricultural Information, 1095–1099, 2008

[3] Nanseki Teruaki, Sugahara Koji, Fukatsu Tokihiro: "Farming Operation Automatic Recognition System with RFID", Japanese Society of Agricultural Informatics, 16(3), 132–140, 2007

[4] Homma Koki, Okai Hisashi et al. "Application of water budget equation model to farmer's fields of 'Tambaguro' soybean", Crop Science Society of Japan (55), 27–32, 2010

Industrial Engineering and Management Science – Lee (Ed)

An improved 4H-SiC MESFET structure with stepped source field plate

HuJun Jia, ZheLin Sun & XiaoYan Pei
Key Lab of Wide Band Gap Semiconductor Material and Device, College of Microelectronics, Xidian University, Xi'an, China

ABSTRACT: A novel 4H-SiC MESFET structure with stepped source field plate (SSFP-MESFET) is proposed. The stepped source field plate consists of upper and lower parts. Two addition field peaks formed at both upper and lower source field plate edges weaken the equipotential lines crowding at the gate edge and enhance the breakdown voltage. The effect of the SSFP on breakdown voltage (V_{BR}), saturation current and RF characteristics are detailed investigated by 2D numerical simulation. The simulated results show that obvious improvement can be obtained for the SSFP-MESEFT compared to the conventional MESFET structure (C-MESFET). To be specific, the V_{BR} of the SSFP-MESFET is enhanced from 72 V to 150 V which means more than 108% improvement is achieved and the output power density (P_{max}) of is about 110% larger. Meanwhile, due to the depletion layer formed under the stepped source field, the gate-drain capacitance of the SSFP-MESFET structure is 23% smaller.

Keywords: 4H-SiC MESFET; stepped source field plate; breakdown voltage; gate-drain capacitance

1 INTRODUCTION

Silicon carbide (SiC) metal semiconductor field effect transistor (MESFET) has been a competitive candidate for high power and high frequency applications due to its superior material property, such as wide band gap, high breakdown field, high electron saturation velocity, and large thermal conductivity (Hoon et al. 2006, Zhang et al. 2007, Hossein 2010, 2011, Noorbakhsh et al. 2012, Aminbeidokhti & Orouji 2013). These outstanding characteristics allow 4H-SiC MESEFT to provide significantly higher output power density than the conventional semiconductor materials based devices, which means to obtain the same output power density, 4H-SiC MESFET requires lighter weight and smaller size. As its power and frequency applications, a large breakdown voltage is required for 4H-SiC MESFET while maintaining a high drain current. However, these requirements are contradictory to each other: A high drain current needs a large product of channel thickness and doping concentration ($t_C \times N$), while to achieve a large breakdown voltage, the channel doping concentration is required to be not too large. Meanwhile, a thick channel will reduce the ratio of gate length to channel thickness (L_G/t_C), which will lead to a drain induced bias lowering (DIBL) effect and cause a decrease of device performance as well as circuit performance (Zhua et al. 2006, Hossein & Orouji 2012).

In order to overcome this drawback, for the first time, an improved 4H-SiC MESFET structure with stepped source field plate is proposed (SSFP-MESFET) in this paper. The stepped source field plate (SSFP) modifies the surface electric field of the device by the depletion layer formed underneath. Two introduced addition field peaks weaken the lines crowding at the gate edge and enhance the breakdown voltage. On the other hand, the stepped source field plate is electrically connected to the source operating like shield electrode between gate and drain which avoid the increase of the gate-drain capacitance (C_{GD}). The capacitance induced by the stepped source field becomes the source-drain capacity (C_{DS}) which can be absorbed in the output tuning network. Moreover, the depletion layer induced by the source field plate modulates the electric near the gate edge and decreases the gate-drain capacitance. The fabrication of the structure can be realized completely free of ion implantation which avoid the implantation-induced crystal damages (Noorbakhsh et al. 2012).

2 SSFP-MESFET AND SIMULATION METHOD

Figure 1 (a) and (b) shows the schematic cross sections of the C-MESFET and the SSFP-MESFET. The device structure consists of layers: a semi-insulating substrate, a p-buffer layer with its thickness of 0.55 μm, doping concentration of 4×10^{15} cm^{-3}, a 0.30 μm thickness, 2.79×10^{17} cm^{-3} doping level n-type channel layer, and a heavily dopped n-cap layer with its doping concentration of more than 10^{20} cm^{-3}. There is a dielectric 450 nm thick Si_3N_4 film on the surface, a stepped field plate connected to the source terminal is settled on the Si3N4 film extended out over the gate to the gate-drain region. The stepped field plate

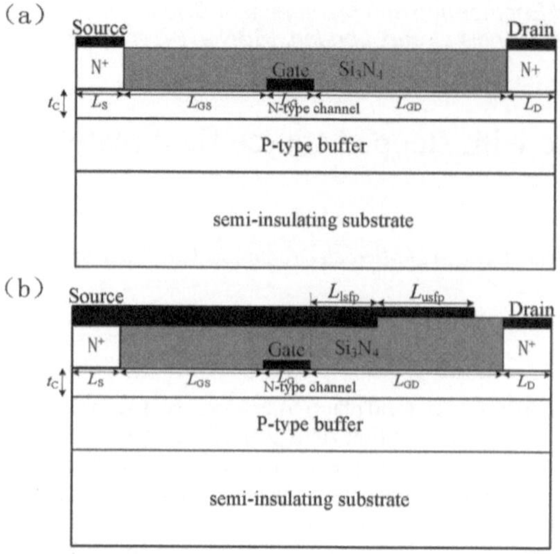

Figure 1. Schematic cross sections of (a) the C-MESFET and (b) the SSFP-MESFET.

consists of lower and upper parts, the length of the lower part $L_{\mathrm{lsfp}} = 0.7\,\mu\mathrm{m}$, the length of the upper part $L_{\mathrm{usfp}} = 0.9\,\mu\mathrm{m}$, the height of the step between the lower and upper part is $0.1\,\mu\mathrm{m}$. The other dimensions of the device are gate length $L_G = 0.5\,\mu\mathrm{m}$, source length and drain length $L_S = L_D = 0.5\,\mu\mathrm{m}$, gate-source spacing $L_{GS} = 1.5\,\mu\mathrm{m}$, gate-drain spacing $L_{GD} = 2\,\mu\mathrm{m}$. The parameter of the C-MESFET is the same as the SSFP-MESFET except otherwise to have a meaning for comparison.

The 2D numerical simulation is realized with TCAD from ISE (Integrated Systems Engineering 2004). To obtain a precisely simulated result, except for basic Poisson, drift/diffusion equations and Fermi, the different physical models are considered such as Thermionic, Doping Dependence, High Field Saturation and Carrier Scattering for mobility, Avalanche (Eparallel) for generation, Band2Band for recombination and Incomplete Ionization for incomplete ionization of dopants.

3 RESULT AND DISCUSSION

Figure 2 illustrates that the saturation current (I_{Dsat}) of the SSFP-MESEFT is slightly smaller than that of the C-MESEFT. The decrease on I_{Dsat} is totally due to the depletion layer formed by stepped source filed plate leading to a narrower effective channel thickness. It should be noted that the dc output conductance of the SSFP-MESFET is better than the one of C-MESEFT obviously. This is due to the stepped source field plate enhancing the effect of the vertical electric field and suppressing the dependence of the drain current with drain voltage.

Figure 3 shows the three-terminal breakdown characteristics of two structures under bias condition $V_{GS} = V_T$, respectively. As can be seen, the breakdown

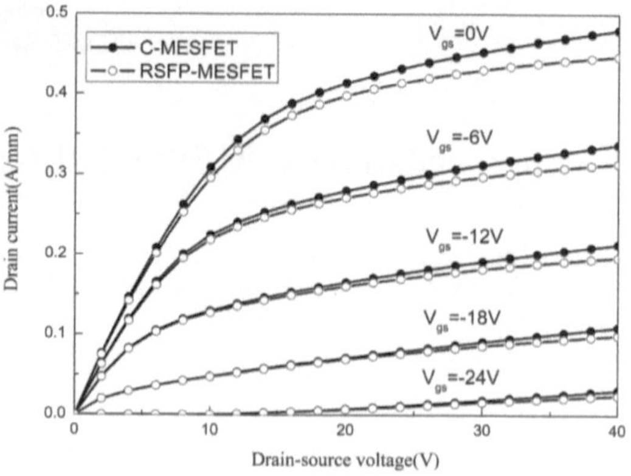

Figure 2. Simulated output characteristics of the C-MESFET and the SSFP-MESFET.

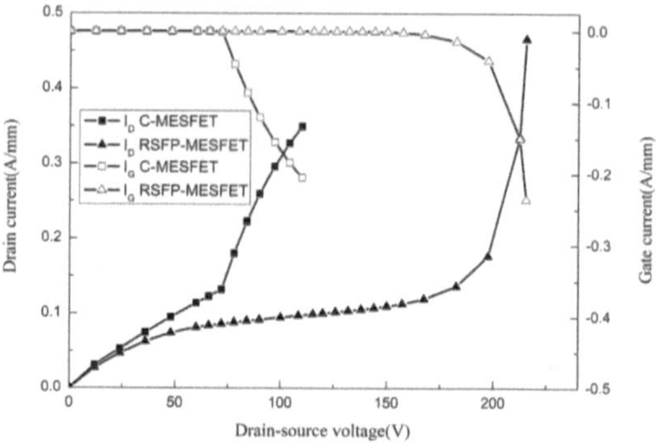

Figure 3. Simulated three-terminal breakdown characteristics of the C-MESEFT and the SSFP-MESFET.

voltage V_{BR} of the SSFP-MESFET is 150 V compared to 72 V of the C-MESFET, which means more than 108% promotion obtained. The figure also indicates that the increased drain current meanly comes from the leakage current at the gate electrode. It illustrates the avalanche occurs between the gate and the drain instead of between the source and the drain. Further investigation turns out that the breakdown happened at the gate edge towards the drain. The equipotential contours of the C-MESFET and SSFP-MESFET are shown in figure 4 (a) and (b), respectively. From the figure, due to the stepped source field plate, it shows a talent effect on spreading the potential lines towards the drain. Two addition field peaks as is the equipotential crowding in the figure at both upper and lower source field plate edges nearby the drain weaken the lines crowding at the gate edge and enhance the breakdown voltage.

The great increment on breakdown voltage with slight decrement on saturation drain current leads to a significant improvement on maximum output power density. An Class A maximum theoretical output power density can be calculated as follow (Ladbrooke 1989):

$$P_{\max} = \frac{I_{\mathrm{Dsat}}(V_{BR} - V_{\mathrm{knee}})}{8} \tag{1}$$

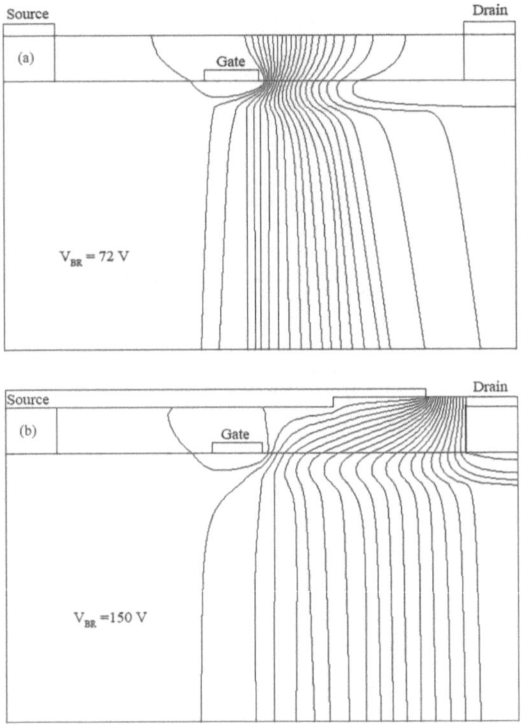

Figure 4. Equipotential contours of (a) the conventional 4H-SiC MESFET and (b) the proposed 4H-SiC MESFET with stepped source field plate at breakdown voltage respectively.

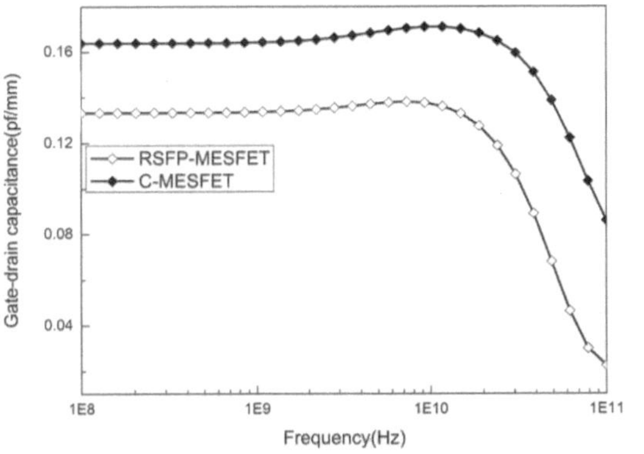

Figure 5. Simulated gate-drain capacitance as a function of frequency for the SSFP-MESFET and C-MESFET at $V_{GS} = -10$ V, $V_{DS} = 40$ V.

where the V_{knee} is the knee voltage. The calculated P_{max} for C-MESFET and SSFP-MESFET is 1.9 w/mm and 4.0 w/mm, it is more than 110% improvement. Hence the proposed structure has a superior DC performance than the convention one.

Figure 5 illustrates that the gate-drain capacitance (C_{GD}) of RFSP-MESFET and C-MESFET, respectively. The bias conduction are $V_{GS} = -10$ V, $V_{DS} = 40$ V. It is obvious from the figure that the gate-drain capacitance of the SSFP-MESFET is smaller than that of the C-MESFET. The better gate-drain capacitance of the RFSP-MESFET benefits from the depletion layer in the vicinity of the gate edge induced by the stepped field plate. The capacitance can be computed by the following equation:

$$C = \varepsilon_s \frac{A}{d} \tag{2}$$

where ε_s is semiconductor permitivity, A is the area of the capacitance plate, d is the distance between capacitance plates. In semiconductor devices, d is considered as the distance of the metals and the depletion around them (Aminbeidokhti & Orouji 2011). The additional depletion introduced by the stepped source field enlarges the distance i.e. d, therefore, the gate-drain capacitance is reduced. It should be noted that with the the gate narrowing, the draining of depletion at gate edge extension will become dominating in the gate-drain capacitance. Meanwhile, the SSFP-MESFET structure has a lager output resistance, as discussed previously. Therefore, the SSFP-MESFET with a better power gain can be predicted.

4 CONCLUSION

In this paper, an improved 4H-SiC MESFET structure with stepped source field plate is proposed (SSFP-MESFET). The DC and RF characteristics are simulated and compared with the conventional MESFET(C-MESFET) by 2D numerical simulation. By employing two addition field concentration the breakdown voltage enhance from 72 V of the C-MESFET to 150 V of the SSFP-MESFET, which means over 108% increment is obtained on voltage. The saturation drain current is slightly lowering compared to the C-MESFET. The maximum theoretic output density improve more than 110% form 4.0 w/mm of the LSFP-MESFET to 1.9 w/mm for that of the proposed structure. Meanwhile, the gate-drain capacitance is 23% smaller than that of the C-MESFET, the output resistance is lager than the C-MESFET, as a result, the proposed structure has a better power gain. The superior performance shows that the proposed SSFP-MESFET has a great potential for high frequency and high power applications.

ACKNOWLEDGEMENT

This work was supported by National Key Basic Research Program of China (973 Program) under grant No.2014 CC339900.

REFERENCES

Aminbeidokhti Amirhossein & Orouji Ali A. 2011. A novel 4H-SiC MESFET with modified channel depletion region for high power and high frequency applications. *Physica E: Low-dimensional Systems and Nanostructures* 44(3): 708–7113.

Aminbeidokhti Amirhossein; Orouji Ali A. 2013. A new double-recessed 4H-SiC MESFET with superior RF characteristics. *International Journal of Electronics* 100(2): 171–179.

Hoon Joo Naa et al. 2006. Fabrication and characterization of 4H-SiC planar MESFETs. *Microelectronic Engineering* 83(1): 160–164.

Hossein Elahipanah. 2010. Simulation and optimization of high breakdown double-recessed 4H-SiC MESFET with metal plate termination technique. *Superlattices & microstruct* 48(6): 529–540.

Hossein Elahipanah. 2011. Record gain at 3.1 GHz of 4H-SiC high power RF MESFET. *Microelectronics Journal* 42(2): 299–304.

Hossein Elahipanah & Orouji Ali A. 2012. Orouji. Gain improvement and microwave operation of 4H-SiC MESFET with a new recessed metal ring structure. *Microelectronics Journal* 43(7): 466–472.

Integrated Systems Engineering. 2004. manual for DESSIS, ISE TCAD Release 10.0.

Ladbrooke P.H. 1989. FETs in amplifiers in: *MMIC design: GaAs FET's and HEMT's*: 259–321. Artech House, Northwood, MA.

Noorbakhsh et al. 2012. Improved 4H-SiC MESFET with double source field plate structures. *Superlattices & Microstructures* 51(5): 553–562.

Zhua C.L. et al. 2006. Improved performance of SiC MESFETs using double-recessed sructure. *Microelectronic Engineering* 83(1): 92–95.

Zhang Jinping et al. 2007. Improved double-recessed 4H-SiC MESFETs structure with recessed source/drain drift region. *Microelectronic Engineering* 84(12): 2888–2891.

Industrial Engineering and Management Science – Lee (Ed)
© 2015 Taylor & Francis Group, London, ISBN 978-1-138-02647-6

High voltage ride through control strategy of PMSG based on multi-modes control[1]

WanYe Yao & WeiNa Ma
North China Electric Power University, Baoding, China

Kai Bai & Peng Song
North China Electric Power Research Institute, Beijing, China

ABSTRACT: To improve the ability of the high voltage ride through of the direct-driven Permanent Magnet Synchronous Generator Wind Turbines, this paper proposes a high voltage ride through control strategy based on multi-modes control and the operational characteristics of the wind turbines. According to the extent of the grid voltage swells and the DC bus voltage, changing the control mode in the grid side converter via the mode selector contributes to the ability of the high voltage ride through of the wind turbines. PSCAD model simulation results show that this control strategy can not only ensure continuous work of the direct-driven Permanent Magnet Synchronous Generator Wind Turbines during grid voltage swells, but also can effectively improve the reactive power compensation capabilities of wind turbines, which will promote the secure and stable operation of the grid.

1 INTRODUCTION

In recent years, the direct-driven permanent magnet synchronous generator (PMSG) wind turbines have more advantages over the doubly-fed induction generator (DFIG) wind turbines, including the possibility of multi-pole design with a gearless construction that offers slow speed operation and reduces noise, easiness in accomplishing fault-ride through and grid support, etc. Therefore, the penetration of the PMSG gradually increases (Seul, Ki. Kim & Eung, Sang. Kim 2007).

With the continuous expansion of the wind turbine's unit capacity and the wind farm scale, more and more countries develope their own connection rules. Meanwhile, many transmission system operators (TSO) have defined more stringent requirements, which include the requirements of the ability of the low voltage ride through (LVRT) of the wind turbines as well as the ability of the high voltage ride through (HVRT) (Dai, Huizhu & Chi, Yongning 2012). The general shape of the high voltage traces characteristic (USA 2007) developed by Western Electricity Coordinating Council (WECC) has been shaped definitely as noted in Figure 1; Australia is the first country to develop HVRT guideline for grid-connected wind turbines. When the high-voltage side grid voltage surges to 130% of rated voltage, the wind turbines should be maintained 60 ms instead of off-grid and provide a large enough recovery current (Australian 2007), High Voltage Ride-Through Curves illustrated on Figure 1; E.ON's requirements (E.ON Netz 2006) for connection of wind turbines in Germany determined the wind turbines must support the grid voltage with additional reactive current during a voltage swell. The voltage control must take place within 20 ms after fault recognition by providing a reactive current on the low voltage side of the generator transformer amounting to at least 2% of the rated current for each percent of the voltage swell.

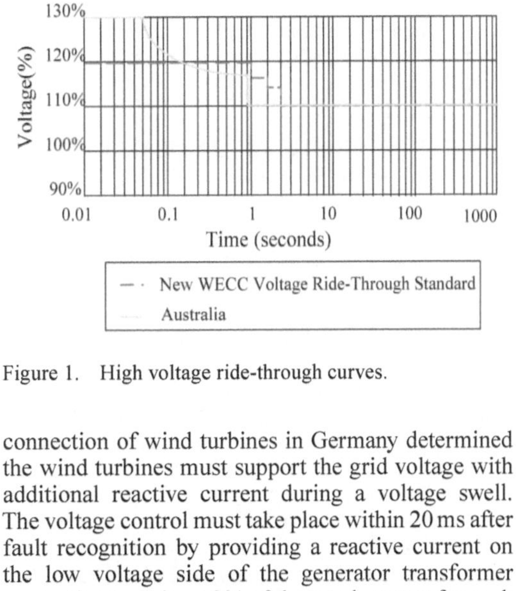

Figure 1. High voltage ride-through curves.

The impact of voltage dip on wind turbines and the low voltage ride through technology are commonly studied currently. However, the impact of grid swell on wind turbines and the corresponding HVRT strategies have not been given sufficient attention. Relative to the HVRT technology (He, Yikang & Hu, Jiabing 2012, M, Mohseni, Masoum, M.A.S. & Islam, S.M. 2011, Xie, Zhen Zhang, Xing & Song, Haihua 2012, Xie, Zhen, Zhang, Xing & Yang, Shuying 2012) of DFIG, the introduction about the HVRT technology of PMSG

[1]Supported by the Fundamental Research Funds for the Central Universities

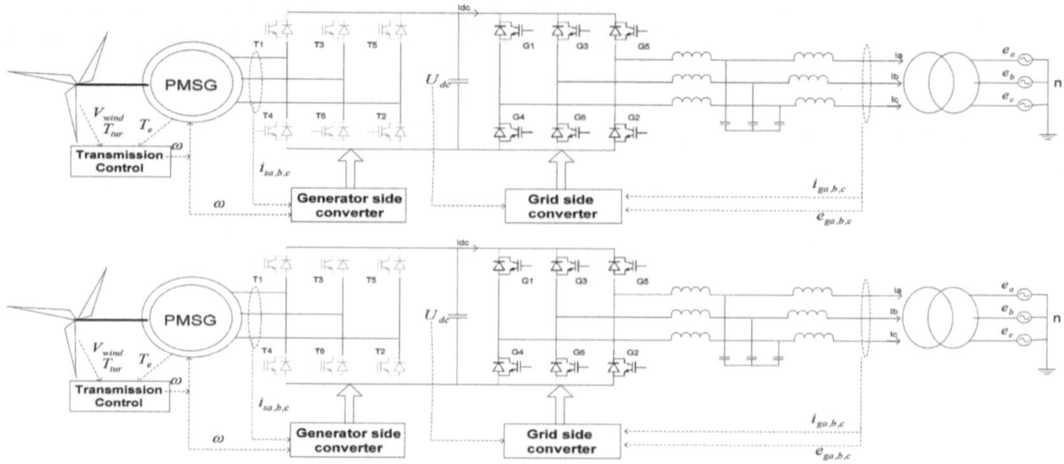

Figure 2. Configuration of PMSG.

is few and far between. Therefore, in-depth study of the operational characteristics of the wind turbines during the voltage swell and proposing a feasible high voltage ride through control strategy are very necessary and urgent.

This paper firstly analyzes the operational characteristics of the PMSG wind turbines during the voltage swell, then discuss the possible method to improve the results, finally proposes a high voltage ride through control strategy of PMSG which based on multi-modes control.

2 CHARACTERISTIC ANALYSIS OF PMSG UNDER GRID VOLTAGE SWELL

In order to analyze the operating characteristics of PMSG under the grid voltage swells (Dai, Huizhu & Chi, Yongning 2012, Seul, Ki. Ki & Eung, Sang. Kim 2007, Gao, Feng & Zhou, Xiaoxin 2011), the establishment of simulation model is necessary. The model topology is shown in Figure 2.

The outputs of wind generator and machine side converter are unaffected by the grid voltage swells, which will cause the power imbalances between wind turbines and grid. Then charging the excess energy into the DC bus capacitor through the grid side converter results in the rise of DC bus voltage . The normal operation of the converter will be threatened, even the wind turbine would be damaged. Grid voltage rising above the protect value of wind turbine causes wind turbine off-grid. Consequently, the stable operation of the grid will be impacted greatly. The substantive control of the inverter is the control of the generator, so the grid side converter control can only be as the main research object without changing the topology of wind turbines (Morinaga, Shuta & Izumi & Yuya 2013, Li, Jianlin & Hu, Shuju & Kong, Deguo 2008, Zhang, Xing & Zhang, Chongwei 2012).

Grid-side converter is controlled by the most widely used method of grid voltage controlling the outer ring and the current controlling the inner. The grid voltage

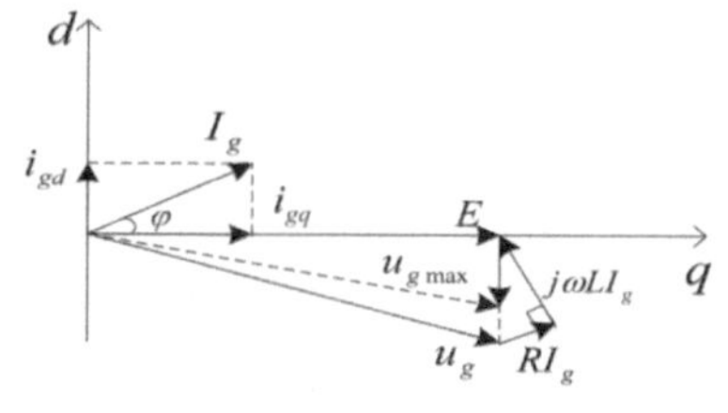

Figure 3. Steady-state voltage space vector of grid-side converter.

is directed to the q-axis, the output active power of grid side converter and the output current to q-axis component of the grid-side converter is proportional, the output reactive power is proportional to the grid-side output current in the d-axis component, so the active and reactive power decoupling control can be achieved.

The steady-state equations of grid-side converter in synchronous rotating coordinate system are:

$$\begin{cases} u_{gd} = -Ri_{gd} + e_d + \omega Li_{gq} \\ u_{gq} = -Ri_{gq} + e_q - \omega Li_{gd} \end{cases} \tag{1}$$

The steady-state voltage space vector of grid-side converter is shown as Figure 3.

According to voltage space vector modulation theory, the modulation ratio m under no over-modulation must satisfy:

$$m = \frac{\sqrt{u_{gd}^2 + u_{gq}^2}}{U_{dc}/2} \leq \frac{2}{\sqrt{3}} \tag{2}$$

Combining the Figure 3 and Equation (2), we can draw: if the power factor angle ϕ is constant, the output impedance of the voltage vector falls necessarily on the hypotenuse of the triangle, and the maximum u_{gmax} is strictly limited by the DC voltage (Morinaga, Shuta & Izumi & Yuya 2013, Xu, Hai-liang & Zhang,

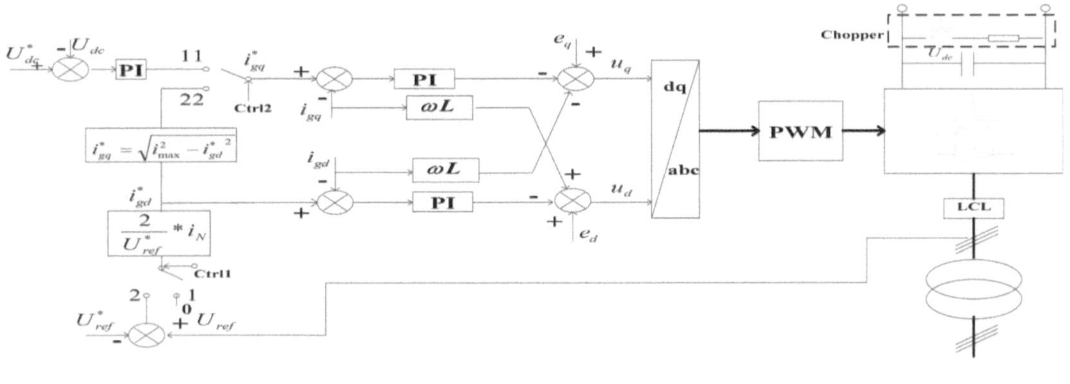

Figure 4. HVRT control strategy.

Wei & Chen, Jiansheng 2013, Yao, Jun & Liao, Yong & Zhuang, Kai 2009).

$e_d = 0$, $e_q = E$ (E is the grid phase voltage peak), combining the formula (1) and (2), simultaneously ignoring the resistance R:

$$U_{dc} \geq \sqrt{3}\sqrt{(E - \omega L i_{gd})^2 + (\omega L i_{gq})^2} \tag{3}$$

Thus when the network side converter operates in the state of unity power factor, the DC voltage U_{dc} should be not less than the grid line voltage peak.

The maximum active power output of wind turbine can be achieved when grid-side converter operating in the state of unity power factor. If grid side converter continues to maintain unity power factor condition under the grid voltage swells circumstances, the voltage limits of the converter will be over and even the converter be damaged. Therefore, certain measures should be taken to prevent the DC bus overvoltage and ensure the normal operation of the wind turbine during the grid voltage swells. Considering the degree of difficulty and cost issues, the strategy of combining chopper unloading circuit and changing the grid-side converter working conditions should be adopted to improve the degree of pressure of the DC bus capacitors and the ability of reactive power compensation of wind turbines.

3 HVRT CONTROL STRATEGY BASED ON MULTI-MODES CONTROL GRID-SIDE CONVERTERS

Wind turbines should have the reactive power compensation ability to suppress the grid voltage swells and the excess energy can be consumed by the chopper unloading circuit on the DC side to prevent wind turbines off the grid during grid voltage swells.

According to the literature (Xu, Hai-liang & Zhang, Wei & Chen, Jiansheng 2013) that the reactive power output of wind turbines is proportional to the magnitude of the grid voltage swells. The specific logical relations can refer to the E.ON HVRT requirements: the voltage control must take place within 20 ms after fault recognition by providing a reactive current on the low voltage side of the generator transformer amounting to at least 2% of the rated current for each percent of the voltage dip. A reactive power output of at least 100% of the rated current must be possible if necessary.

$$i_{gd}^* = 2\frac{U_{ref} - U_{ref}^*}{U_{ref}^*} * i_N \tag{4}$$

where i_{gd} is the reactive current, U_{ref} is the measured grid voltage RMS, U_{ref}^* is the referred grid voltage RMS, i_N is the rated current.

In order to avoid the grid-side converter over current, the active current amplitude is required to satisfy:

$$\left|i_{gq}^*\right| \leq \sqrt{i_{\max}^2 - i_{gd}^{*\,2}} \tag{5}$$

($i_{\max}$ is the maximum allowable current of grid-side converter.)

HVRT control strategy block diagram is shown in Figure 4.

The control strategy utilizes two mode selectors Ctrl1 and Ctrl2 to change the control modes of the grid-side converter, so that the active current reference value and the reactive current reference value can be set to change with the grid voltage. Beside, the Chopper circuit can also be controlled to work according to the DC bus voltage changing though consuming excessive energy to suppress the DC bus voltage rising.

The specific control strategy is: making the poor between the real-time measurements of the grid voltage RMS and grid voltage setting value, the status of which above the limit or not will be set the input signal of Ctrl1. Ctrl1 selector chooses model '1' when the grid voltage is normal to set the reactive current at zero, while Ctrl2 chooses model '11' that grid voltage controlling the outer ring and the current controlling the inner, keeping the wind turbine maintaining the unity power factor status. When grid voltage surges to make the real-time measurements of grid voltage exceeds limit, the mode '2' will be selected. The reference reactive current value i_{gd}^* is calculated through

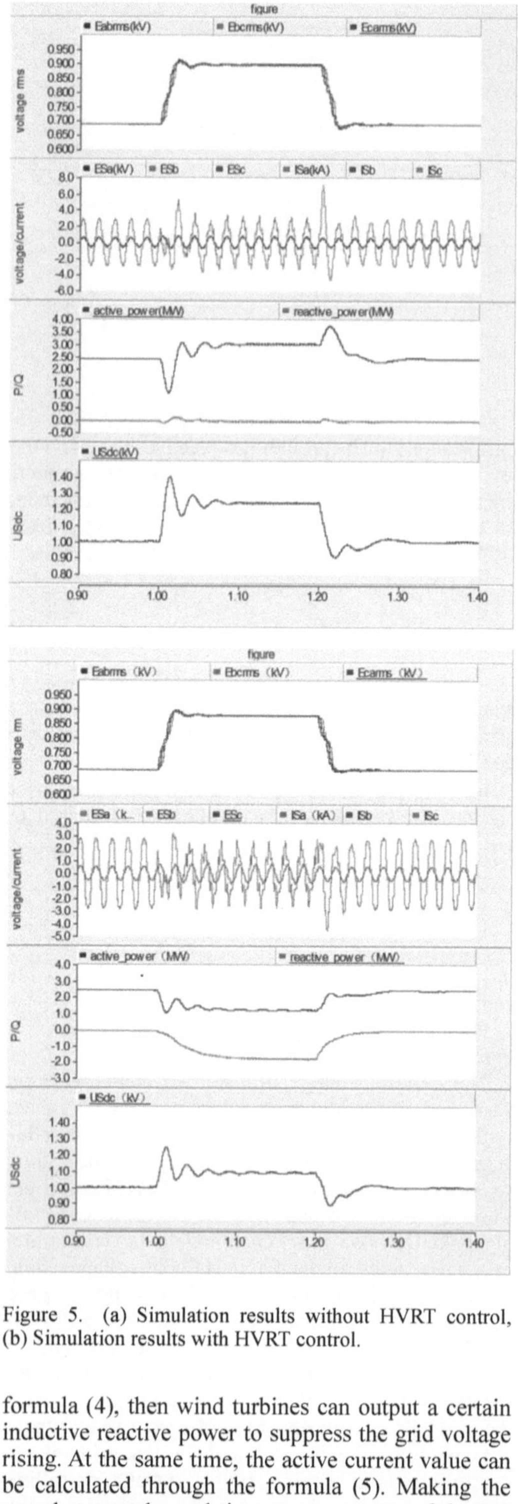

Figure 5. (a) Simulation results without HVRT control, (b) Simulation results with HVRT control.

formula (4), then wind turbines can output a certain inductive reactive power to suppress the grid voltage rising. At the same time, the active current value can be calculated through the formula (5). Making the poor between the real-time measurements of the DC bus voltage value U_{dc} and the DC bus voltage setting value, the status of which above the limit or not will set the input signal of Ctrl2. Ctrl2 selector chooses model '22' when the poor exceeds the limit, so at this time the reactive current increases and the active current slants will be decreased, which indirectly promotes the reactive compensation capacity of wind turbine. Meanwhile the input signal of Ctrl2 is the Chopper circuit switch signal, the Chopper starts to consume the excess energy when the DC bus voltage is higher than the limit, to prevent the DC bus over-voltage and ensure that the wind turbine can be normally operating.

4 SIMULATION VERIFICATION

The simulation model of the 2.5 MW direct-driven Permanent Magnet Synchronous Generator Wind Turbine is built on the PSCAD simulation platform to simulate the actual operation situation. The grid-side converter is connected at 690V level and through a step-up transformer to the 35 kV. The voltage on the low voltage side of the generator transformer rises to 130% of the rated voltage after the wind turbine stably operating for 1.5 s, then returns to the normal after 200 ms. The simulation results are shown in Figure 5.

Comparing the operating characteristics of wind turbine between adopting the HVRT control strategy and not, we can draw: wind turbines can output a certain inductive reactive power through changing the control mode of the grid-side converter during the grid voltage swells. The active power output of wind turbines reducing can contribute to suppressing the degree of grid voltage swells. Chopper circuit is turned on when the DC bus voltage rises above the limit. And the DC bus voltage dips to the allowable range with the excessive energy consumed. As a result, the normal operation of the wind turbine is ensured during the grid voltage swells, so that wind turbine has the ability of HVRT.

5 CONCLUSION

This paper proposes a high voltage ride through control strategy derived from multi-modes control on the basis of the deeply research on the operational characteristics of the wind turbines. According to the extent of the grid voltage swells and the DC bus voltage, changing the control mode in the grid side converter via the mode selector, contributes to the ability of the high voltage ride through of the wind turbines. Control strategies used in this paper are easily combined with existing low voltage ride through control strategies to enable wind turbines to run high and wear low.

REFERENCES

Australian. 2008. Australian energy market commission. *National Electricity Rules*.

Cai, Guoying. 2009. A research of the maximum wind energy tracking strategy of the permanent-magnetic direct-drive wind power. *Xiamen University*.

Dai, Huizhu. & Chi, Yongning. 2012. Comparison study on grid codes for connecting wind farm into power system.*Electric Power* 10: 1-6+11.

E.ON Netz. 2006. Grid code for high and extra high voltage. Bayreuth, *E.ON Netz Gwbh.*

Gao, Feng. & Zhou, Xiaoxin. 2011. Electromechanical transient modeling and simulation of direct-drive wind turbine system with permanent magnet synchronous generator. *Power System Technology* 11: 29–34.

He, Yikang & Hu, Jiabing. 2012. Several hot-spot issues associated with the grid-connected operations of wind-turbine driven doubly fed induction generators. *Proceedings of the CSEE* 32(27): 1–15.

Li, Jianlin, Hu, Shuju. & Kong, Deguo. 2008. Studies on the low voltage ride through capability of fully converted wind turbine with PMSG. Automation of Electric Power Systems 32(19): 92–95.

Mohseni, M., Masoum, M.A.S. & Islam, S.M. 2011. Low and high voltage ride-through of DFIG wind turbines using hybrid current controlled converters. *Electrical Power System Research* 81(7): 1456–1465.

Morinaga, Shuta. Izumi. & Yuya. 2013. Output power control of a PMSG based wind turbine in strong wind conditions. *IEEE International Symposium on Industrial Electronics*.

Seul, Ki. Kim. & Eung, Sang. Kim. 2007. PSCAD/EMTDC-Based modeling and analysis of a gearless variable speed wind turbine. *IEE transactions on energy conversion* 22(2): 421–430.

USA. 2007. The technical basis for the new WECC voltage ride-through (VRT) standard, *WECC Board.*

Xie, Zhen., Zhang, Xing. & Song, Haihua. 2012. Variable damping based control strategy of doubly fed induction generator based wind turbines under grid voltage swell. *Automation of Electric Power Systems* (3): 39–46.

Xie, Zhen., Zhang, Xing. & Yang, Shuying. 2012. High voltage ride-through control strategy of doubly fed induction wind generators based on virtual impedance. *Proceedings of the CSEE* 32(27): 16–23.

Xu, Hai-liang., Zhang, Wei. & Chen, Jiansheng. 2013. A high-voltage ride-through control strategy for DFIG based wind turbines considering dynamic reactive power support. Proceedings of the CSEE.

Yao, Jun., Liao, Yong. & Zhuang, Kai. 2009. A low voltage ride – through control strategy of permanent magnet direct- driven wind turbine under grid faults. *Automation of Electric Power Systems* 33(12): 91–96.

Zhang, Xing. & Zhang, Chongwei. 2012. *PWM rectifier and its control*. Beijing: China Machine Press.

Industrial Engineering and Management Science – Lee (Ed)
© 2015 Taylor & Francis Group, London, ISBN 978-1-138-02647-6

Design of a 1.5GSPS 5bit folding and interpolating ADC with distributed S/H folding amplifiers in 90 nm CMOS technology

ShiLiang Li, Hui Hong & Shuai Liu
Institute of Microelectronic CAD, Hangzhou Electronic University, Hangzhou, China

ABSTRACT: The Folding and Interpolating Analog to Digital Converters (F&I ADCs) with sampling rate higher than 1GSPS are widely used in high speed data acquisition, military and avionics imaging, etc. This paper presents a 1.5GSPS 5bit ADC which is implemented by using folding and interpolating ADC architecture with only 11 comparators needed. To eliminate the frequency multiplied effect while expanding the input full scale range, fully differential sampling and holding (S/H) amplifiers are proposed to implement the folding amplifiers. The F&I ADC is designed in 1.2 V 90 nm CMOS technology. The simulation results show the sampling rates can achieve as high as 1.5 GSPS, with the DNL, INL of less than 0.5 LSB, the SNR of 31 dB and the ENOB of 4.8 bits. The power dissipation is about 600 mW, and the FOM of the ADC is 12 pJ/conversion-step.

1 INTRODUCTION

Very high speed analog to digital converters (ADCs) with sampling rate which higher than 1GSPS are widely used in software defined radio (SDR), high speed oscilloscopes, wireless communications testers, etc. Such a high speed ADC can be implemented by using full flash ADC architecture, or pipelined ADC architecture or time-interleaved architecture, but the most popular one is folding and interpolating ADC architecture. A 5bits full flash ADC using conventional architecture in Li's work needs at least $2^5 - 1$ or 31 comparators, which cost a large chip area and consume a great power dissipation. The conventional continuous-time folding amplifiers utilized in Hai's and K's works also has frequency multiplication problem, which deteriorated the performance of folding amplifiers.

The designed ADC in this paper needs only 11 comparators. And by utilizing the switched S/H folding amplifiers to replace the conventional continuous time amplifiers, the folding amplifiers can work in a higher frequency. Because the succeeding analog pre-processing circuitries are now operating in the discrete-time domain, the inherent frequency multiplication induced by the folding operation of the succeeding folding amplifiers is eliminated.

2 SYSTEM CONSIDERATION

The proposed F&I ADC consists of two sub-ADCs: a 3 bit coarse ADC and a 2 bit fine ADC. This resolutions configuration of the coarse sub-ADC and the fine sub-ADC can simplify the design of folding amplifiers and encoders. The MSB 3 bit codes are generated by comparing the sampled and held signal with the reference voltages Vref2 ~ Vref8 paralleled. The comparing results which are 7bit thermometer codes are converted to 3 bits natural codes by the proposed fat tree encoder.

Meanwhile, the 9 reference voltages Vref1 ~ Vref9 are sent to the folding amplifiers in the 2 bit fine ADC. The folding amplifiers are design to had a minimum 3X folding rate, because a 3X folding rate can guarantee the linear region of those folding amplifiers not overlapped too much, so the folding amplifiers can still work in a narrow input full scale range. The 9 reference voltages and the sampled signal are folded by 3 groups of 3x folding amplifiers. The frequency of the folded signal is multiplied by conventional continuous time folding amplifiers, and the proposed switched folding amplifiers which have the function of sampling and holding are used. The signal after S/H can be considered as a DC signal, so the frequency multiplication effects won't influence the performance of the folding amplifiers.

Three groups of 4X resistors interpolating network are utilized to generate extra zero-crossing points. To implement a 5 bit ADC, at least 2^5 or 32 zero-crossing points are needed. The 3 folded signals are interpolated by the 4X resistors interpolating networks, and a total 9×4 or 36 zero-crossing points are generated, which can meet the demand of the 5bit ADC. The 12 signals are folded again by the following 4 groups 3X folding amplifiers.

The four groups twice folded signals are sent to 4 zero-crossing detectors and the cycling thermometer codes generated by the zero-crossing detectors are converted to natural codes by the followed XOR-OR encoder. The 3 bit MSB codes converted by the coarse ADC and the 2 bit LSB codes and 1bit synchronizing

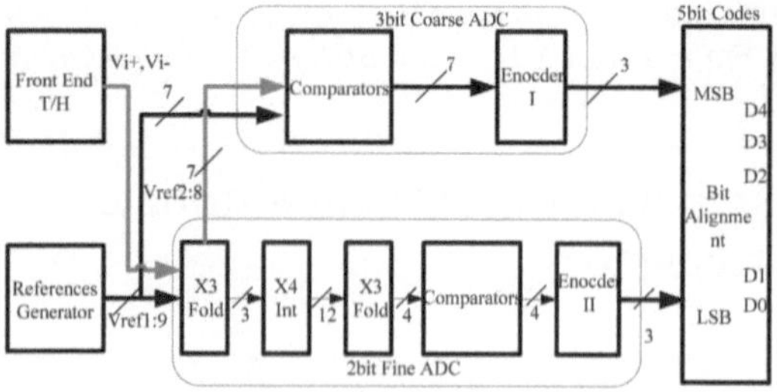

Figure 1. Block diagram of the designed 5 bit F&I ADC.

code are re-aligned by the bits alignment module to get the final 5bit converted codes. The block diagram of the 5bit F&I ADC is illustrated in figure 1.

3 CIRCUIT IMPLEMENTATION

3.1 *Full differential S/H 3X folding amplifiers*

Folding amplifiers are the basic and the key circuit modules in F&I ADC. The 3x folding amplifiers used in the fine 2 bit ADC can be implemented by using common amplifiers. We used a NMOS input differential pair rather than a PMOS pair, because the mobility of a NMOSFET is larger than a PMOSFET, so a NMOSFET can have a small size with the same transconduction Gm, i.e., the same bandwidth. In figure 2 (a), resistors are used as the loads, so the widest bandwidth can be achieved. Considering the bandwidth, the gain and the input/output range of the amplifier, the bias tail current sink is designed with a value of 4 mA and the value of the load resistors is 150 Ω. Then the output common mode voltage is 0.9 V.

To obtain a fully differential form of the input signal and the reference voltages, a switched differential input architecture is proposed as shown in figure 2(a). The switches S1 and S2 are working in orthogonal phases as shown in figure 2(b). In the first phase, the switches S1 are turned on while the switches S2 are turned off, the potentials of the input gates are Vg1 = Vg2 = VCM, and the output potentials are Vop = Von = 900 mV. At this phase, the capacitors Cs are charged and the total charges are Qt = (VCM-Vref) × Cs. In the next phase, the switches S1 are turned off while the switches S2 are turned on, the plates of Cs which are connected to the input gates are floated, so the charges stored in Cs kept unchanged. The other plates are connected to Vip (Vin), so the potentials of the plates connected to the gates are Vg1 = VCM + Vip − Vrp, Vg2 = VCM + Vin-Vrn. The difference of Vg1 and Vg2 is Vg1 − Vg2 = (Vip − Vin) − (Vrp − Vrn) = Vi − Vref, so we can get the relationship between Vi and Vref from the relationship between Vg1 and Vg2. The 3X fully differential folding amplifiers can be implemented by paralleling the outputs of three these amplifiers.

Due to the small voltage gain of these amplifiers, the linear amplify regions of the folding amplifiers are overlapped seriously, which are unwanted. The way to solve the linear region overlap problem is cascading some stages of preamplifiers in front of the folding amplifiers.

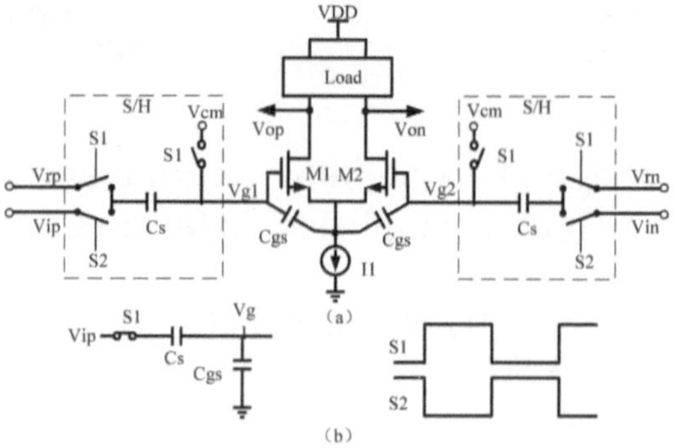

Figure 2. Fully differential S/H input amplifier for folding amplifiers.

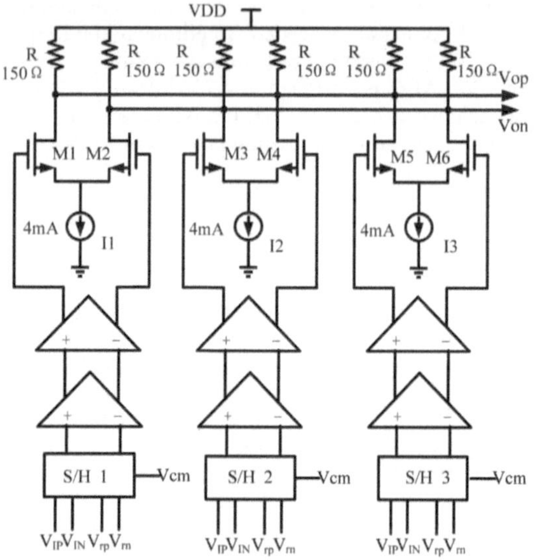

Figure 3. 3X folding amplifiers.

3.2 *4X resistors interpolating network*

Interpolating network is used to generate extra zero-crossing points. The conventional 4X resistors interpolating network is illustrated in figure 4(a). The extra 3 zero-crossing points X1, X2, X3 between F1 and F2 are generated by the 4X resistors interpolating network. Ideally the extra zero-crossing points are distributed uniformly. The X2 is generated by F1 and F2, and due to the same phase of F1 and F2, X2 is ideally in the middle of F1 and F2. However, X2 don't have the same phase with F1 and F2, so the zero-crossing points X1 and X2 which generated by F1 and X2, F2 and X2 respectively, are not in their ideal positions, causing zero-crossing points drift. The zero-crossing drifts would deteriorate the DNL errors seriously, and influence the performance of the whole ADC system.

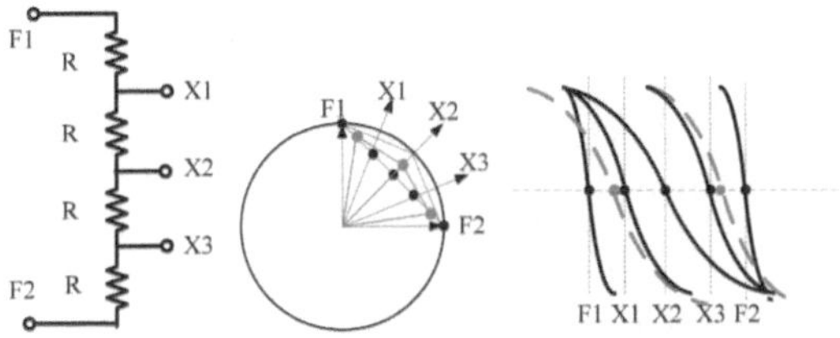

(a) Conventional 4x resistors interpolating network

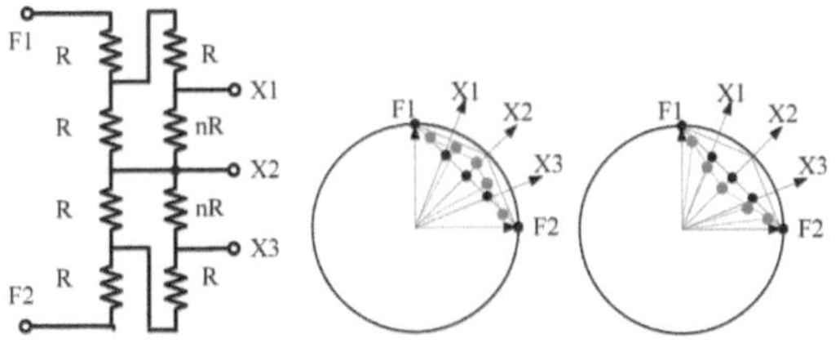

(b) Proposed 4X resistor interpolating network

Figure 4. 4X resistors interpolating network architecture.

A proposed 4x resistors interpolating network is illustrated in figure 4(b). Two resistors which have the ratio 1:n are connected between the original X1, X3 and X2, and the nodes connected R and nR generate the new zero-crossing points X1 and X3. The zero-crossing drifts problem can be solved by tuning the ratio 1:n. By using this structure, the DNL errors are reduced effectively, so the performance of the whole ADC is improved. In order to not affect the performance of the folding amplifiers, the total values of the interpolating resistors are proposed be a slight smaller than the output resistors of the folding amplifiers.

3.3 *Zero-crossing detect comparator*

The comparators cascading folding amplifiers are working as detectors to detect whether zero-crossing point changed when the input signal changed. To meet the requirement of a high speed sampling rate, the comparator should have a very short setting time, a very small transport delay time and a big voltage slew rate. For a 1.5GSPS sampling rate, the period time of the clock is about 666 ps, so the total delay time of the comparator should be smaller than half a clock period, or 333 ps. To meet the high speed requirement, higher speed comparator architecture is proposed as the figure 5 shown. The comparator consists of two pre-amplifier and a latch. The high speed characteristic is due to the positive feedback which exists in the latch and the sample comparator architecture.

In the first phase, the switches controlled are turned on by S2 and the M1 is turned off by S1. In the phase, no static current flows in the latch. The potentials of the node A and node B are following the outputs of the preamplifiers cascaded. In the next phase, the switches are turned off by S2 and M1 is turned on by S1. The current source M2 provides a large instantaneous bias current. If the initial potential at node A is large than node B, the instantaneous bias current and the positive feedback exist in the latch pushes node A to the VDD

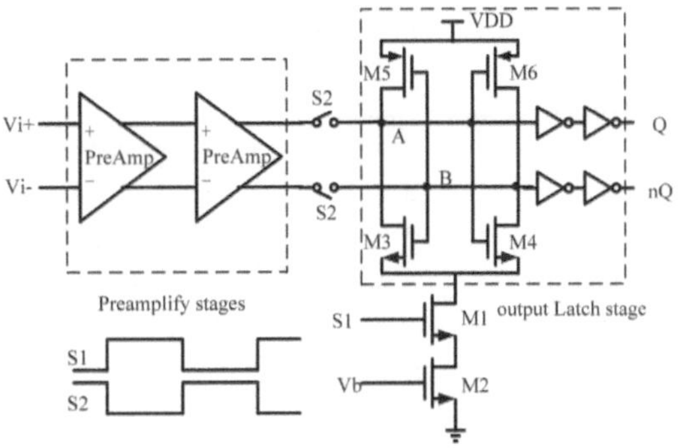

Figure 5. The proposed high speed zero-crossing detect comparator.

rail and pulls A to the GND rail and keeps this state until next comparing phase. No large static current is flowing in the latch, so the comparator is also energy-saving. The input full scale range of the 5 bit F&I is differential 800 mV, so 1LSB = 800 mV/32 = 25 mV. A common 8 mV input offset voltage of the comparator will not cause the so called bubble codes, which are harmful for encoding the thermometer codes to the natural codes.

3.4 *Encoders and bit alignment*

After the operation of comparing, two types of codes existed in the F&I ADC. One is 7 bit thermometer codes generated by the 7 comparators in the 3 bit coarse ADC, the other is 4 bit cycling thermometer codes in the 2 bit fine ADC. There are 8 states in 7 bit thermometer codes, so they can be encoded into 3 bit natural codes; there are also 8 stages in 4 bit cycling thermometer codes, so 3 bit natural codes are obtained as well. However, the LSB of the 3 bit codes in the coarse ADC and the MSB of the 3 bit codes in the fine ADC are the same, so the total resolution of the ADC is 5 bit.

The 7 bit thermometer codes are encoded by using the fat tree encoder. The small delay time of this encoder makes it suitable for high speed operation. The logic expression of the fat tree encoder is illustrated in (1).

$$\begin{aligned} &MSB = T4 \\ &MSB-1 = T2\cdot\overline{T4} + T6 \\ &MSB-2 = (T1\oplus T2\oplus T3)\cdot\overline{T4} \\ &+(T5\oplus T6\oplus T7) \end{aligned} \tag{1}$$

The 4 bit thermometer codes are encoded by using the XOR-OR encoder. The logic expression of the XOR-OR encoder is illustrated in (2).

$$\begin{aligned} &(LSB+2) = A \\ &(LSB+1) = (A\oplus B)+(B\oplus C) \\ &LSB = (A\oplus B)+(C\oplus D) \end{aligned} \tag{2}$$

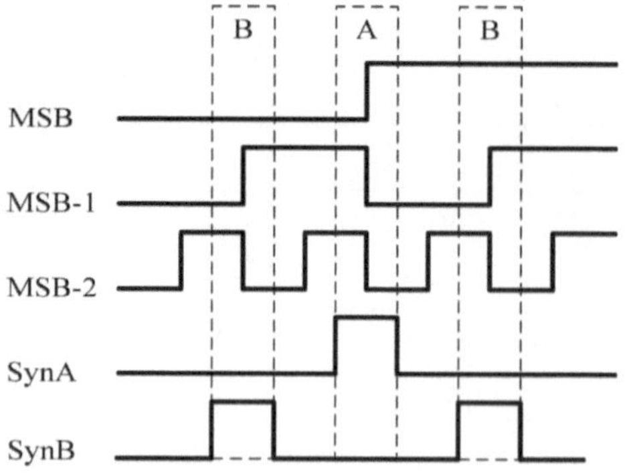

Figure 6. Implementation of bit alignment.

Although the coarse ADC and the fine ADC is working simultaneously, but they are two independent ADCs, the converted results should be re-aligned to complete 5 bit codes. The bit alignment is implemented by using the following logic operation shown in (3).

$$\begin{aligned} &MSB' = SynA \cdot \overline{(LSB+2)} + MSB \\ &(MSB-1)' = SynA \cdot (LSB+2) \\ &+ SynB \cdot \overline{(LSB+2)'} + (MSB-1) \\ &(MSB-2)' = LSB+2 \end{aligned} \quad (3)$$

The synchronizing signal SynA and SynB are given as figure 10 shown and the simulation results before and after bit alignment are illustrated in figure 6.

4 SIMULATION RESULTS

The overall 5 bit F&I ADC is implemented by using 90 nm CMOS technology and verified by using the CADENCE specter simulation tools.

With a ramp input simulation, the output codes of the designed ADC are collected and the output codes with an ideal DAC are reconstructed. The DNL and INL are calculated with the reconstructed ramp signal. The reconstructed ramp signals without bit alignment are shown in figure 7. No missing code is found in the ADC with bits alignment.

A 31.25 MHz sinusoidal wave is input to the ADC at a sampling rate 1 GSPS and an ideal DAC is also used to reconstructed the output codes of the ADC as shown in figure 8. A 32 points FFT is applied to the reconstructed wave, and the major dynamic characteristics of the designed ADC are obtained. The 32 points spectrum is shown in figure 9. The calculated results show that the SFDR of the designed ADC is 31.2 dB, and the SNR is 31 dB.

5 CONCLUSION

A 1.5 GSPS 5 bit ADC which is implemented by using folding and interpolating ADC architecture is proposed in this paper. By utilizing the switched S/H folding amplifiers to replace the conventional continuous time amplifiers, the inherent frequency multiplication induced by the folding operation of the succeeding folding amplifiers is eliminated, so the folding amplifiers can work in a higher frequency.

The simulated results show that the 5 bit F&I ADC can operating at a 1.5 GSPS sampling rate. The calculated SNR is 31 dB and the corresponding ENOB

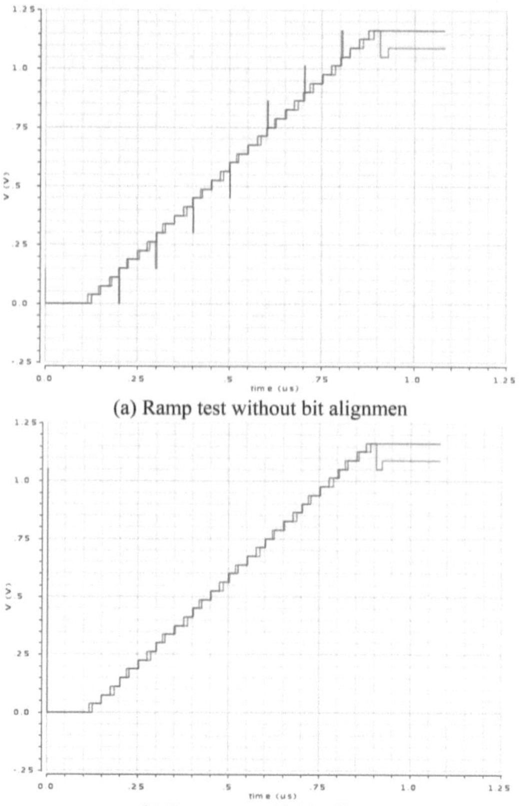

(a) Ramp test without bit alignmen

(b) Ramp test with bit alignment

Figure 7. Ramp test for ADC without bit alignment (a) and with bit alignment (b).

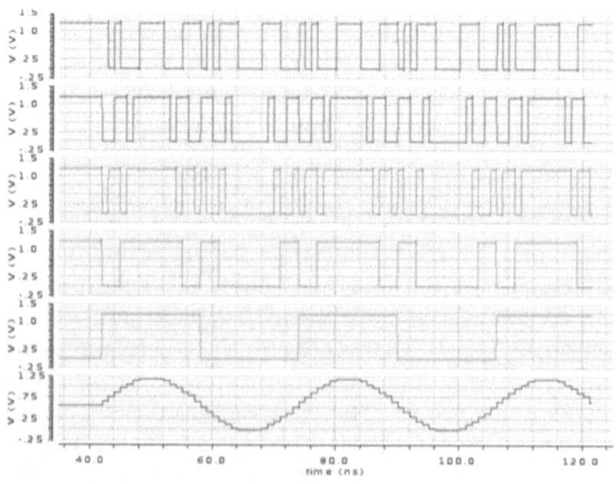

Figure 8. The reconstructed sin-wave.

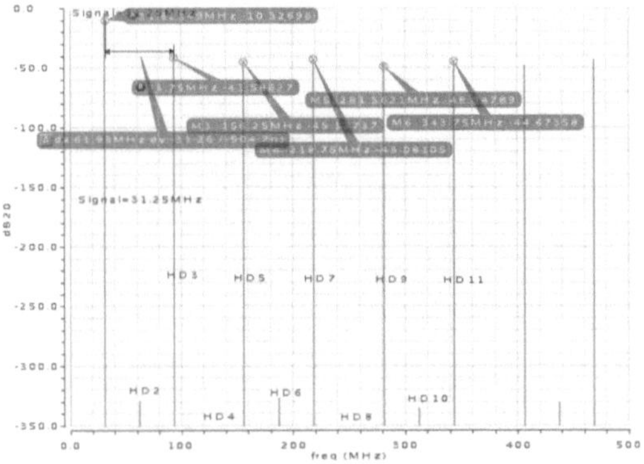

Figure 9. 32 points FFT of the reconstructed sin-wave.

Table 1. Summary and comparison of the performance of the designed ADC.

	This work	Li shi wen's	Hai rong, Y's	E. Zhian Tabasy's
Technology	CMOS 90 nm	CMOS 90 nm	CMOS 90 nm	CMOS 90 nm
Architecture	F&I*	Flash	F&I	T&I* SAR
Sampling Rate (Gsps)	1.5	1.0	1.25	1.6
Supply voltage (V)	1.2	1.2	1.0	1.3
Input range (mV)	800	940	–	–
ENOB (bit)	4.8	4.0	7.0	4.75
DNL (LSB)	±0.5	0.28	–	+0.86/−0.51
INL (LSB)	±0.5	0.14	–	+1.1/−2.6
SNR (dB)	31.00	21.00	48.10	30.35
SFDR (dB)	31.2	–	47.5	–
Power (mW)	600	–	207	20
FOM (pJ/c)	12	–	1.2	0.46
Years	2014	2012	2008	2012

*F&I is short for Folding & Interpolating.
*T&I is short for Time Interleaved.

is 4.8 bit and no missing codes are found. The total power dissipation of the whole ADC is about 600 mW, and the calculated FOM is 12.5 pJ/conversion-step.

ACKNOWLEDGMENTS

In this paper, the research is sponsored by the Nature Science Foundation of China (Project No. 61107025) and the Key Innovation Team Project of Zhejiang Province (Project No. 2010R50010).

REFERENCES

Ali, A.M., Morgan, A., Dillon, C., Patterson, G., Puckett, S., Hensley, M., Stop, R., Bhoraskar, P., Bardsley, S., Lattimore, D., Bray, J., Speir, C. and Sneed, R. A 16b 250MS/s IF-Sampling Pipelined A/D Converter with Background Calibration, *In Proc. ISSCC 2010 , 2010*: 292–294.

Daegyu Lee, Jincheol Yoo, Kyusun Choi, and Jahan Ghaznavi. Fat tree encoder design for ultra-high speed flash A/D converters, *Circuits and Systems, 2002. MWSCAS- 2002. The 2002 45th Midwest Symposium on, 2008*: 87–90.

Hai rong, Y. and Chang, M.C.F. A 1-V 1.25-GS/s 8-Bit Self-Calibrated Flash ADC in 90-nm Digital CMOS, *Circuits and Systems II: Express Briefs, IEEE Transactions on, 2008*: 668–672.

Hui Pan and Asad A. Abidi. Signal Folding in A/D Converters, *Circuits and Systems I: Regular Papers, IEEE Transactions on, 2004*: 3–14.

Li Shiwen, Dang Hua, Gao Peng, et al., Green Computing and Communications (GreenCom), *2013 IEEE and Internet of Things (iThings/CPSCom), IEEE International Conference on and IEEE Cyber, Physical and Social Computing, 2013*: 1829–1833.

Peetz, B., Hamilton, B. and Kang, J. An 8-bit 250 megasample per second analog-to-digital converter: Operation without a sample and hold, *IEEE J. Solid-State Circuits, 1986*: 997–1002.

Shruti, K. and Devashrayee, N.M. Low power folding and Interpolating ADC Using 0.18 um Technology, *International Advances in Computer Engineering Conference, 2010*: 54–58.

Stepanovic, D. and Nikolic, B. A 2.8 GS/s 44.6 mW Time Interleaved ADC Achieving 50.9 dB SNDR and 3dB Effective Resolution Bandwidth of 1.5 GHz in 65 nm CMOS, *VLSI Circuits (VLSIC), 2012 Symposium on, 2012*: 84–85.

Vorenkamp, P. and Roover, R. A 12b 50M Samplels Cascaded Folding and Interpolating ADC, *Circuits and Systems II: Express Briefs, IEEE Transactions on, 2008*: 668–672.

Zhen Liu, Song Jia, Lijiu Ji and Xing Zhang, Low-Power CMOS Fully-Folding ADC with a Novel Bit Synchronization Architecture, *Solid-State and Integrated Circuit Technology, 2006. ICSICT '06. 8th International Conference on, 2006*: 1652–1654.

Zhian Tabasy, E., Shafik, A., Huang, S., Yang, N. Hoyos, S. and Palermo, S. A 6b 1.6GS/s ADC with Redundant Cycle 1-Tap Embedded DFE in 90nm CMOS, *Custom Integrated Circuits Conference (CICC), 2012 IEEE, 2012*: 1–4.

Section 3: Performance evaluation

Industrial Engineering and Management Science – Lee (Ed)
© 2015 Taylor & Francis Group, London, ISBN 978-1-138-02647-6

Impact of emotional intelligence towards occupational stress

Choi Sang Long & Mardhiah Yaacob
Faculty of Management, Universiti Teknologi Malaysia, Malaysia

ABSTRACT: Stress in the workplace reduces employees' productivity and affects their work performance. Emotional intelligence comes to rescue and guides employees to respond appropriately to different stressors. Emotional Intelligence helps to cope up with stressful situations and emotional control behaviour. Emotional intelligence improves work performance by enabling people to manage their emotions. This will be useful especially dealing with stress and perform well under pressure.

1 INTRODUCTION

Emotional intelligence was coined by Salovey and Mayer (1990), which is a way of recognizing that emotions and emotional information were crucial for problem solving and adaptation in everyday life. Salovey and Mayer (1990) defined emotional intelligence as the ability to monitor one's own and other's emotion, to discriminate among them, and to use the information to guide one's thinking and actions.

Stress is the destructive physical, mental and emotional reaction that transpires when there is a poor match between job demands and competencies to manage with job pressure (Saddam, 2010). Occupational stress occurs because of the mismatch between the situation's demands and the individual's coping abilities (AbuAlRub, 2004). Occupational stress is an unpleasant situation that employee experience when the requirement of work-related or not related cannot be counter balance with the ability to resolve them (Halkos and Bousinakis, 2010).

Slaski and Cartwright (2003) explain the relationship between emotional intelligence and stress. They argued that it is the negative emotions and stress that resulted in dysfunctional relationship between aspects of the self and the environment. This also affect the ability to 'read' and manage emotions in one' self and others. Employees should be trained to identify, handle and utilize their emotions to eliminate the hindrances in the way of choosing and advancing their career prospect (Carmeli, 2003). A person should have lower withdrawal intention if equipped with high emotional intelligence. This is due to the ability to better regulate the emotions. Therefore, emotional intelligence enables an employee to exhibit the skills necessary to be aware of and regulate the emotions of themselves as well as the emotions of others. This ability allows an employee to adapt to environment demands and self-perceived emotionally. The use of emotional intelligence is predictors of affect within work environment at the workplace (Kafetsios and Zampatekis, 2008).

The intention of this study is to review the past literature on the impact of emotional intelligence towards occupational stress.

2 THE IMPORTANCE OF EMOTIONAL INTELLIGENCE

Emotional intelligence is important to everyday life because it predict behavior and important outcomes. Individual who are poor at dealing with emotions will have worse relationship, poor mental health, and less career success (Ciarrochi et al., 2001). Ciarrochi et al. (2001) claims that people who can succeed depend on how they deal effectively with emotions and use the emotions to enrich their life. According to Salovey and Mayer (1990), emotional intelligence is combination of individual's ability to adapt to life changes through the use of both rational and emotional coping skills with regards to stress. Therefore, emotional intelligence enables an employee to exhibit the skills necessary to be aware of and regulate the emotions of themselves as well as the emotions of others. This ability allows an employee to adapt to environment demands and be self-perceived emotionally. The use of emotional intelligence is predictors of affect within work environment at the workplace (Kafetsios and Zampatekis, 2008).

According to Salovey and Mayer (1990), emotional intelligence abilities can also positively influence problem solving skills. They argue that positive emotion and moods could change the way in which an individual approach a problem, organize thoughts, process information, apply creative thinking and fulfill other life task. Emotional intelligence also contributed significantly to attitudes to change in the organization, suggesting that the use of an emotional intelligence measure can add significant value within organizations (Vakola et al., 2004).

The emotional intelligence employees able to the proper handle of the negative feelings in a way to express it positively, allowing people to interact and work together without friction (Saddam, 2010). This ability facilitates the employee to redirect their unconstructive stressful reactions and impulses. This is even more crucial particularly if such employee is a leader of the organization. An emotional intelligence employee is a consistent and dedicated employee, open to everything new, even in the most uncertain prospects and sudden changes such an employee tend to perceive as a new opportunity rather than as a threat to personal safety and suffer in stress (Saddam, 2010).

Therefore, emotional intelligence play important role to create the abilities in an employee's to better control the stress in the workplace. The emotional intelligence generates the skill in employee to choose various courses of action to deal stress. Thus, dealing stress without collapsing and positively solving problems can be expected. Commonly, a person feel that one can control the situation if such skill is mastered (Slaski and Cartwright, 2002).

3 EMOTIONAL INTELLIGENCE AND OCCUPATIONAL STRESS

Previous studies reveal that the emotional intelligence individuals have strong attitude to deal stressful events in a positive way (Gardner, 2005; Ciarrochi et al., 2001). Each individual have their own way in dealing with stress. Emotional intelligence is one of the most effective way especially in workplace. The dimension of emotional intelligence is based on individual's understanding and managing of one's own personal emotions as a way to reduce tension and stress (Goleman, 1998).

It has been suggested that emotional intelligence plays an important role in helping individuals to control and manage the negative emotions appropriately and also moderate both the appraisal and experience of stress (Slaski and Cartwright, 2003). It would seem that individual differences in emotional intelligence relating to differences in an individual's ability to appraise his or her own emotions and others emotions. This type of individuals tend to be more open to internal experience and better able to label and communicate those experiences to others (Mayer and Salovey, 1993). Slaski and Cartwright (2003) stressed that individual who have highly emotional intelligence report significantly less stress and experience better physical and psychological health than less emotional intelligence individuals. It can be said that the link between emotional intelligence and stress is founded on the notion that negative emotion and stress are the result of some dysfunctional relationship between aspects of the self and the environment. Negative emotion will lead not only to failure in team effort but also destruction of individual character.

In studies exploring the relationship between emotional intelligence and occupational stress, Bar-On et al. (2000) investigated the associations between the variables in different occupational groups. Bar-On et al. (2000) investigated emotional intelligence in two occupational groups including police officers and paraprofessional personnel in mental health and child care professions. The results of this study indicated that police officers had higher levels of emotional intelligence than the other group. Also, Bar-On et al. (2000) suggested that the police officers are more aware of their own feelings and understand themselves; consequently they can be more adaptable to stressful events, and choose better coping strategies.

A study by Nikolaou and Tsaousis (2002) found a negative correlation between EI and stress at work among the mental health professionals. Nikolaou and Tsaousis (2002) claims that employees with high emotional intelligence confront low occupational stress in their job environment. Employees with low emotional intelligence found to have less self-awareness, and as a result they are not able to cope with their feelings and have high stress which, in turn, has a negative influence on their job satisfaction. Such employees will become unproductive with adverse behavior. Oginska-Bulik (2005) explored the relationship between emotional intelligence and perceived stress in workplace and health related consequences in human service workers. The outcome of the study confirms that emotional intelligence influences occupational stress and prevents employees of human services workers from negative health outcomes. This study also concluded that the ability to effectively deal with emotions and emotional information in the workplace assists employees in coping with occupational stress. Long et al. (2014) further stated that occupational stress affect work performance if left unchecked.

In another study by Landa et al. (2008), it showed that the nurses with high emotional clarity and emotional repair had less levels of stress, yet those with high emotional attention had greater levels of stress. Employee with high emotional attention will find themselves depressed if they sense lack of attention on their appearance at workplace. Ismail, Suh-Suh, Ajis and Dollah (2009) conducted a study to examine the effect of emotional intelligence in the relationship between occupational stress and job performance. The outcome of the study stated that relationship between occupational stress and emotional intelligence significantly correlated with job performance. Statistically, the results confirmed that the emotional intelligence as a mediator to the relationship between occupational stress and job performance.

The study done by Chhabra and Chhabra (2012) to understand the relationship between emotional intelligence and occupational stress found a negative correlation between those two variables among the Indian Border Security Force personnel. There is considerable evidence that the police personnel are suffering from high level of stress taking a heavy toll on the personnel, physically and psychologically. Individuals who can adapt to the situational stress and have better adaptation are more satisfied and less stressful. This adaptability depends on many factors, one of the most important being emotional intelligence which is an effective tool and can help in coping with stress.

4 CONCLUSION

It seems obvious that the employees' ability to effectively deal with emotions assist them in managing their stress in the workplace. The relationship that existed between emotional intelligence and occupational stress could be understood from the perspective that employees with high emotional intelligence are able to monitor their own and others' emotion, discriminate among them and use the information to guide their thinking and actions. They also have the ability to facilitate emotion and cognitive activities such as thinking and problem solving. Therefore, it is important for organizations to put effort and a worthy investment in increasing skills in emotional intelligence among their employees.

ACKNOWLEDGEMENT

Authors wish to acknowledge the Malaysian Ministry of Higher Education and Universiti Teknologi Malaysia under the Research Grant (Vot. 4F349) for supporting and sponsoring this publication.

REFERENCES

Abu Al Rub, R.F. (2004). Job Stress, Job Performance, and Social Support among Hospital Nurses. *Journal of Nursing Scholarship, 36*(1), 73–78.

Bar-On, R., Brown, J.M., Kirkcaldy, B.D., and Thome, E.P. (2000). Emotional expression and implications for occupational stress; an application of the Emotional Quotient Inventory (EQ-i). *Personality and Individual Differences, 28*, 1107–1118.

Carmeli, A. (2003). The Relationship between Emotional Intelligence and Work Attitudes, Behavior and Outcomes: An Examination among Senior Managers. *Journal of Managerial Psychology, 18*(8), 788–813.

Ciarrochi, Chan, Caputi and Roberts (2001). Measuring emotional intelligence in Ciarrochi, Forgas, and Mayer (2001). *Emotional intelligence in everyday life: a scientific inquiry*. 98-132. Phidalephia, Psychology Press.

Gardner, L. (2005). *Emotional Intelligence and Occupational Stress*. PhD Thesis: Swinburne University.

Goleman, D. (1998). *Working with emotional intelligence*. New York: Bantam Books.

Halkos, G. and Bousinakis, D. (2010). The effect of stress and satisfaction on productivity. *International Journal of Productivity and Performance Management*, 59(5), 415–431.

Kafetsios, K. and Zampetakis, L.A. (2008). Emotional intelligence and job satisfaction: testing the mediatory role of positive and negative affect at work. *Personality and Individual Differences, 44*, 712–722.

Long, C.S., Kowang, T.O., Ping, T.A., and Muthuveloo, R. (2014). Investigation on the impact of job stressors on nurses in Malaysia. *Asian Social Science*, 10(4), 67–77.

Mayer, J.D. and Salovey, P. (1993). The intelligence of emotional intelligence. *Intelligence, 17*, 433–442.

Mayer, J.D. and Salovey, P. (1997). What emotional intelligence? In P. Salovey and D. Sluyter (Eds.). *Emotional development and emotional intelligence* (pp. 3–31). New York: Basic Books.

Nikolaou, I. and Tsaousis, I. (2002). Emotional intelligence in the workplace: exploring its effects on occupational stress and organizational commitment. *The International Journal of Organizational Analysis, 10*(4), 327–342.

Oginska-Bulik, N. (2005). Emotional intelligence in the workplace: exploring its effects on occupational stress and health outcomes in human service workers. *International Journal of Occupational Medicine and Environmental Health, 18*(2), 167–175.

Saddam, H.R. (2010). Emotional intelligence and stress: An analytical study of Pakistan banks. *International Journal of Trade, Economic and Finance, 1*(2), 194–199.

Salovey, P., and Mayer, J.D. (1990). Emotional intelligence. *Imagination, Cognition, and Personality, 9*, 185–211.

Slaski, M. and Cartwright, S. (2002). Health, performance and emotional intelligence: An exploratory study of retail managers. *Stress and Health, 18*, 63–68.

Slaski, M. and Cartwright, S. (2003). Emotional intelligence training and its implications for stress, health and performance. *Stress and Health, 19*, 233–239.

Vakola, M., I. Tsaousis and I. Nikolaou. 2004. "The role of emotional intelligence and personality variables on attitudes toward organizational change." *Journal of Managerial Psychology 19*(2): 88–110.

Industrial Engineering and Management Science – Lee (Ed)
© 2015 Taylor & Francis Group, London, ISBN 978-1-138-02647-6

Structural impact of international oil price on China stock market

ChunHong Li, YunQin Zhang & DongWu Wang
Department of International Business, Guangzhou Institute of Technology, Guangzhou, China

HuiHui Gao
Department of Registry, Guangzhou Institute of Technology, Guangzhou, China

ABSTRACT: This paper takes the data from 1996:1 to 2009:12 as our research sample to deeply investigate the structural impact of international oil price on China's stock market. We use econometric models to resolve the decomposition of structural international oil price fluctuations, which includes Supply shocks, Economic demands shocks and Preventive buy shocks (the oil price fluctuations caused by the change of international oil market supply, the economic needs of the international oil price fluctuations and volatility in international oil prices caused by Preventive buy action). Empirical analysis showed that the impact of international oil price fluctuations on China's Shanghai and Shenzhen stock index were complex. The impact of international oil price volatility on China stock market are the joint action results of the three structured international oil fluctuations factors.

Keywords: International oil price; China stock index; structural international oil price fluctuations

1 INTRODUCTION

Oil is highly associated with the national economy and has the reputation of "industry blood". Crude oil is the most important energy and basic raw materials for the smooth operation of national economy. The international community has always been very concerned about oil price volatility on economic activities. The statistical results of Hamilton (2005)[1] Balke (2008)[2] showed that 9/10 of the U.S. recession were after the dramatically rising of oil price which occurred during the period from 1946 to 1973.

Oil price has high volatility and its impact on developing countries can not be ignored. Annualized price volatility for crude oil is approximately 25% per year while natural gas volatility is approximately 40% per year (Fusaro, 1998[3]). Meanwhile the trade off between risk and return is one of the central issues faced by individuals who trade equities, manage portfolios, or engage in capital budgeting. Consequently, the price of crude oil is likely to be a source of risk for stock returns. This concern is particularly relevant in developing energy consumer countries whose import dependence and net import are getting rise with time passing, such as China. Although the economic impact of oil price volatility has its macroeconomic performance, as a micro-scope, its impact begins with affecting the behavior and performance of micro-level such as its impact on the share price in stock market.

2 LITERATURE REVIEW

Following the major oil price shocks of the 1970s, a large amount of researchs had been done on the impact of oil price on economic activities. From the investmental point of view, energy crisis had given big shock on the capital market, but little research had been done about the impact of the oil price on the resource allocation in capital market. Recent contributions found significant effects of oil price shocks on economic activity for most countries in their research samples. As the important role of oil price playing in economic activity, it is natural to expect that oil price shocks may well have impacts on real output inflation and reflect in stock market.

Although economists agreed that oil price changes are an important incentive for the stock price volatility, but in concrete and empirical process, they didn't reach consensus, or even has two contrary views. Chen, Roll and Ross (1986)[4] thought that oil price fluctuations did not affect the asset pricing and the impact of oil price volatility on the stock market is not significant. In contrast, Jones and Kaul (1996)[5] found that oil price increases in the post war period had a significantly detrimental effect on aggregate stock returns. Sadorsky (1999)[6] and Evangelia Papapetrou (2001)[7] denied the conclusions of Chen, Roll and Ross (1986)[4]. They confirmed the significant impact of oil price on the stock market and reported that oil price increases had significantly negative impacts on U.S. stocks and that the magnitude of the effect may have increased since the mid 1980s. Ciner (2002)[8] concluded that a statistically significant relationship existed between real stock returns and oil price futures, but the connection was nonlinear. Park and Ratti (2008)[9] found oil price shocks had a negative impact on stock markets in importer country while Norway as an oil exporter showed a positive response of stock market to the rise of oil price. The study results of Jouko Rautava (2004)[10] implied that Russian economy was influenced significantly by the fluctuations of oil price and the real exchange rate through both long-run and short-term impacts equilibrium conditions.

With the continuing study in-depth, it is impossible to explain and investigate the complex relationship between the crude oil market and economic activity by taking oil price as an exogenous variable. Krey (2007)[11] studied the impact of uncertainties in the crude oil price volatility on the crude oil supply structure and its interactions with oil demand. To specifically study on the relationship between stock market and oil price due to the changes of supply and demand in Crude oil market, Kilian and Park (2009)[12] took the U.S. stock market as research sample and divided specifically the cause of oil price volatility into three factors which included supply factors, demand factors and preventive needs for oil. They found that the price rise due to supply shock had negative correlation with the U.S. share price while the price rise due to demand shock and preventive needs shock had positive correlation with the U.S. share price. Gogineni (2007)[13], Yurtsever and Zahor (2007)[14] confirmed the conclusions of Kilian and Park (2009)[12] statistically.

3 DESCRIPTION OF STRUCTURAL RECOGNITION MODEL OF INTERNATIONAL OIL PRICE FLUCTUATIONS

Dry bulk sea transport level of activity is used on behalf of the world reflects the degree of economic prosperity. Although crude oil is a special product with scarcity, but the basic rule is still subject to fluctuations in the crude oil market supply and demand and influence factors. Then the research framework of Kilian and Park (2009) [12] is adopted to separate driving factor behind the international oil price fluctuations affecting China Asymmetry stock market volatility in international oil prices heap. Its dynamic impact on China's stock market is then differentiated. Given the minimalist style VAR model can not directly identify the structural impact, Sims (1986) [15] identified that the use of long-term structural impact of recursive VAR model will be developed into SVAR models. SVAR models use three kinds of drivers decomposed fluctuations in international oil prices. The establishment of a global monthly oil production includes natural logarithmic value DOSP, actual shipping worldwide monthly changes in the value of the natural logarithm of the index DWE and the natural logarithm of the difference in the world of international oil prices after the monthly three variables SVAR model SOP data, as following:

$$A_0 Y_t = \alpha + \sum_{i=1}^{j} A_i Y_{t-i} + u_t \tag{1}$$

wherein $Y_t = (DOSP_t, DWE_t, SOP_t)$, and u_t^{demand}, u_t^{supply}, $u_t^{spec\_demand}$ and $u_t = (u_t^{supply}, u_t^{demand}, u_t^{spec\_demand})$, represent oil supply impact, economic demand impact and of preventive purchase impact, all of them are white noise, sequences are orthogonal, and covariance is 0. Model Selection month lag period, three variables through unit root test, *DOSP*, *DWE* and *SOP* are stationary time series in crude oil production, changing with the behavior of the global economy, international oil price fluctuations in SVAR decomposition. By minimalist style itself VAR residual vector $\varepsilon_t = (\varepsilon_t^{supply}, \varepsilon_t^{demand}, \varepsilon_t^{spec\_demand})'$ gets the solve of vector SVAR impact u_t. ε_t means oil production, the global economic situation and the actual international oil price fluctuations disturbance, the disturbance from the structural impact of the economic system, that is a combination of oil supply impact, economic demand impact, and preventive purchase impact. The correlation of three impacts ε_t can be described by $\varepsilon_t = Cu_t$. In the meanwhile, the transformation matrix is

$$C = \begin{bmatrix} c_{11} & c_{12} & c_{13} \\ c_{21} & c_{22} & c_{23} \\ c_{31} & c_{32} & c_{33} \end{bmatrix}$$

4 INDEX SELECTION AND DATA ACQUISITION

According to Kilian's (2009)[12] research methods, this paper uses global oil production as the natural logarithm of monthly changes in the value DOSP, actual shipping worldwide monthly changes in the value of the natural logarithm of the index DWE monthly data on the number and nature of the difference in the world after oil prices SOP's three-variable SVAR model. The impact of fluctuations in international oil prices will decompose in order to facilitate comparative analysis of the impact of international oil price fluctuations due to different reasons for China's stock market. Dry bulk sea transport of the level of activity of various primary products and raw materials can reflect the extent of the world's economic prosperity. When the index was up significantly, the world economy and trade and prosperity, on behalf of a large range of world economic growth, and vice versa. Global maritime dry bulk freight index has a direct impact on the demand for oil consumption. Global dry bulk freight index is based on a number of nautical traditions of dry bulk cargo routes than the price of a reorganization of the structure of the composite index in the shipping

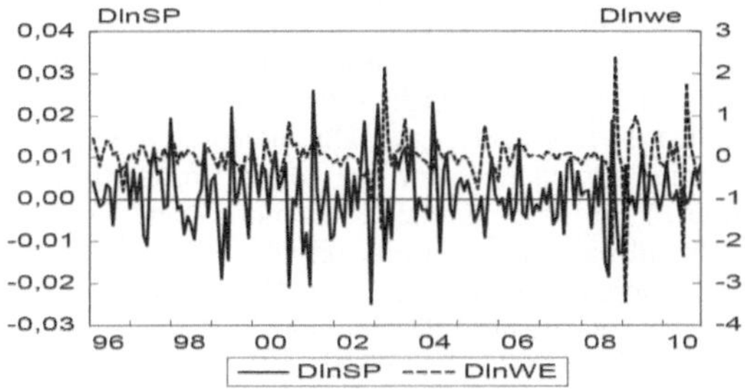

Figure 1. Logarithm difference of world oil production and world economic.

Table 1. ADF unit root test Results of related variables.

Series	ADF Statistics	Difference sequence	ADF Statistics
Ln(*SP*)	−1.277	dLn(*SP*)	−11.840***
Ln(*WE*)	−2.462	dLn(*WE*)	−10.653***
Ln(*SOP*)	−3.948	d Ln(*SOP*)	−15.344***

***indicates a significant level of 1%.

market. Although the impact of the global dry bulk freight index will be sailing weather, war and environmental changes and other emergencies, the impact of these factors on the global dry bulk freight index sailing is just short and unexpected, which does not affect its behavior on behalf of the world economy effectiveness.

5 UNIT ROOT TESTS

SVAR model requires variables in the system are stationary time series, and it is necessary to model the three-variable unit root test, in order to avoid generating a pseudo-variable regression of non-stationary concrete test. Results are shown in Table 1. International oil prices on the Shanghai and Shenzhen stock index closing sequence number exists no unit root so that null hypothesis is rejected. Difference sequence performed its stationary test, and in a significant level of 1%, the original series are I (1) process. Test type (c, t, p), where c represents a constant term, t represents the trend item, P represents the number of lags, the lag order EVIEWS6.0 selected automatically by the software according to AIC Akaike information criterion; wherein a significant level of ADF %, 5% and 10% respectively of the critical −3.467, −2.878 and −2.575.

ADF test results show that changes in the value of the international supply of crude oil, the world economy and the international oil price movements in the index series are not stationary sequence of three variables at the 1% level of significance, but in the 10% significance level still can not reject the null hypothesis, non-stationary sequence. Further changes in the value of the world's oil supply, first-order differential changes in the world economy and the international oil price index sequences can be tested, at

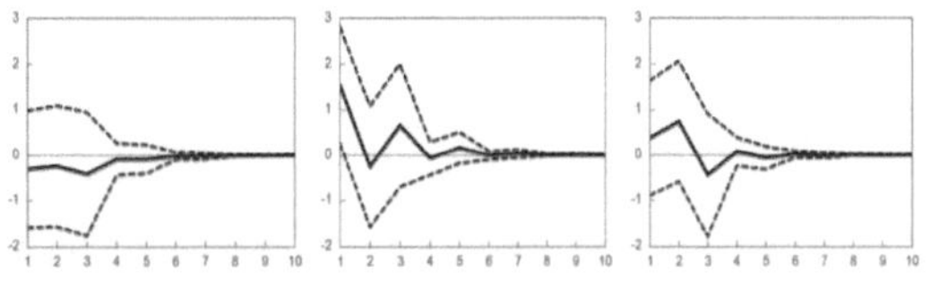

Figure 2. Impact of structural oil price on Shanghai stock returns.

the 1% significance level to reject the null hypothesis, indicating that the three variables by differential is stationary time series. In order to avoid a reduction in the observed value, and ultimately ensure the SVAR model range spans, all variables consistent three-variable time-series interval span 1996:2-2010:12. Follow-up studies will adopt changes in the value of world production of crude oil from February 1996 to December 2010, the world economy and the international oil price fluctuations in the value of the first-order differential (i.e., the international oil price fluctuations) to build three-variable SVAR model for the study of international crude oil price volatility structural decomposition.

6 STRUCTURAL IMPACT OF OIL PRICE FLUCTUATIONS ON SHANGHAI AND SHENZHEN STOCK MARKETS

Figure 2 shows the effect of supply and demand factors and preventive purchase driven by the international oil price fluctuations on the Shanghai stock index gains concentrated in the short term. Economic needs of the international oil price fluctuations and volatility in international oil prices preventive purchase gain a rising on the Shanghai stock index positive impact, the international oil price fluctuations on the supply of income households have a negative effect to the stock shock. But the international supply of oil price fluctuations on the Shanghai stock index returns the smallest impact on the larger factors driven by the precautionary buying volatility of international oil price volatility on the Shanghai stock index gains impulse response. Demand generated by economic factors and preventive factors driving the purchase of fluctuations in international oil prices. The oil supply fails to replenish the short term due to the increase in oil demand will boost international oil prices, increasing business costs, reduce investment, reduce stock market stock index gains; while international oil supply shocks occur, the international oil price decreases, reducing the cost of doing business, but because of the presence of wage rigidity, reduced amount of the increase in investment leads to slight gains of the stock market index.

Figure 3 is a supply and demand factors and preventive purchase driven fluctuations in international oil prices on the Shenzhen stock index returns. Supply factors caused by the impact of international oil price fluctuations on the Shenzhen stock index gains resulting from the smallest, larger by a factor-driven preventive buy the volatility of international oil price fluctuations on the Shenzhen stock index

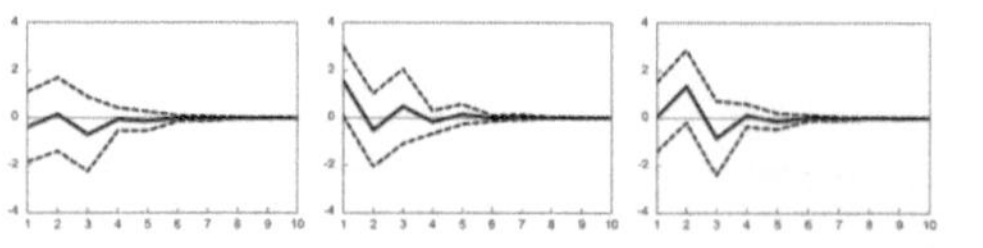

Figure 3. Impact of structural oil price on Shenzhen stock returns.

gains impulse response, demand factors and fluctuations in international oil prices due to supply factors of the impact direction opposite the Shenzhen stock index gains. When the economic needs of the international oil price fluctuations and preventive purchase occurred, failure to replenish the supply of short-term increase in demand for oil will inevitably boost international oil prices, increasing business costs, inhibit investment, reducing the stock index gains; while supplies When international oil price fluctuations, the international oil prices lower, reduce business costs, but the existence of wage rigidity, resulting in less incremental investment, the stock market index gains slightly. There are some differences in the impact of changes in supply, demand factors and preventive factors driving the purchase of international oil price fluctuations on the Shanghai and Shenzhen stock index closed revenue. As the Shanghai stock market opened early, market mechanisms and information disclosure is relatively mature, resulting in the same amount of information that leads to greater impact Shenzhen stock price volatility, the more sensitive response to the Shenzhen stock index information.

In short, as the world changes and the continuous adjustment of industrial structure and upgrade times, considering the impact of research on the economic behavior of international oil price fluctuations, a combination of structural changes is also required in the international crude oil market to have relatively square correct interpretation and analysis, in order to address the impact of the role and impact of international oil price fluctuations caused by economic behavior.

7 CONCLUSION

We use econometric models to resolve the decomposition of international oil price fluctuations structural fluctuations, including Supply shocks, Economic demands shocks and Preventive buy shocks. Empirical analysis of international oil price fluctuations and the role of China's Shanghai and Shenzhen stock index on income and related industry sector index returns. The impact of international oil price volatility on China's stock market is volatile international oil structured by the three factors (including the price fluctuations caused by the change of international oil market supply, the economic needs of the international oil price fluctuations and volatility in international oil prices caused by Preventive buy) the result of joint action.

Despite the lower fuel consumption of China's economic strength and low inflation, economic decline and environmental factors that affect the international supply of oil price volatility on China's economy; economic needs of the international oil price volatility on the Shanghai and Shenzhen stock markets and related industry sector index gradually increased role in promoting our country, for our country promoting the role of the stock market even more than the international oil price volatility on China's stock market inhibition. International oil price fluctuations and supply preventative buy international oil price fluctuations on China's Shanghai and Shenzhen stock index and related industry sectors weaker inhibition.

ACKNOWLEDGMENTS

This work is partially supported by Impact and Analysis of International Oil Price Fluctuation on China Stock Market – 2013 Institute level scientific research project of Guangzhou Institute of technology (No. 201301). The authors also gratefully acknowledge the helpful comments and suggestions of the reviewers.

REFERENCES

[1] Hamilton, J.D. Oil and the Macroeconomy [J]. The New Palgrave Dictionary of Economics, Palgrave Mac Millan Ltd, in S. Durlauf and L. Blume 2005, 2
[2] Balke, N.S., Brown, S.P.A. & Yucel, M.K. Oil Price Shocks and the U.S. Economy: where Does the Asymmetry Originate? [J]. Energy. 2002, 23(3): 27–52
[3] Fusaro, P.C. Energy Risk Management: Hedging Strategies and Instruments for the International Energy Markets [J]. McGraw Hill, New York. 1998,
[4] Chen, N.F., Roll, R. & Ross, S.A. Economic Forces and the Stock Market [J]. Business. 1986, 59(3): 383–403
[5] Kaul, G. & Jones, C.M. Oil and the Stock Markets [J]. Finance. 1996, 51(2): 463–491
[6] Sadorsky, P. Oil Price Shocks and Stock Market Activity [J]. Energy Economies. 1999, 21(5): 449~469
[7] Evangelia Papapetrou. Oil Price Shocks, Stock Market, Economic Activity and Employment in Greece [J]. Energy Economics. 2001, 23(5): 511–532
[8] Ciner, C. Energy Shocks and Financial Markets: Nonlinear Linkages [C]. The MIT Press. 2001, 5(3): 203–212
[9] Jungwook Park & Ronald A. Ratti. Oil price shocks and stock markets in the U.S. and 13 European countries [J]. Energy Economics. 2008, 30(5): 2587~2608
[10] Rautava, J. The role of oil prices and the real exchange rate in Russia's economy [J]. Comparative Economics. 2004, 32(2): 297–314
[11] Krey, V., Martinsen, D. & Wagner, H.J. Effects of Stochastic Energy Prices on Long Term Energy Economic Scenarios [J]. Energy. 2007, 32(12): 2340–2349
[12] Lutz Kilian & Cheolbeom Park. The Impact of Oil Price Shocks on the U.S. Stock Market [J]. International Economic Review. 2009, 50(4): 1267–1287
[13] Gogineni, S. Oil and the Stock Market: An Industry Level Analysis [J]. The Financial Review. 2010, 45(4): 995–1010
[14] Puah Chin-Hong, Tan Lay-Phin, Md Isa & Abu Hassan. The Nexus between Oil Price and Stock Performance of Power Industry in Malaysia [J]. The 11th International Business Research Conference, Sydney, Australia, 2–4, December 2009, 1–18

Industrial Engineering and Management Science – Lee (Ed)
© 2015 Taylor & Francis Group, London, ISBN 978-1-138-02647-6

Detecting human error symptom of body movement in monotonous work

Yohei Tontani, Yusuke Kajiwara & Hiromitsu Shimakawa
Ritsumeikan University, Japan

ABSTRACT: Factories experience problems such as accidents and product defects. Accidents cause injuries to workers, and sometimes cost them their lives in the worst case. Product defects cause vendors to loose customer trust. They also increase production cost because of delays in production processes. Main causes of these problems are human errors. This paper focuses on the movement of workers who repeat same tasks in line production. When they work well, their movement is similar in every repeated task. We assume they have good rhythm in those cases. The paper proposes a method to calculate the rhythm of a worker and the effectiveness of the method is proven by an experiment.

1 INTRODUCTION

Factories experience problems such as accidents and product defects. Accidents cause injuries to workers, and sometimes cost them their lives in the worst case. Product defects drain customer confidence in vendors. They also increase production cost because of delays in production processes. One of the major causes for these problems is human errors. The introduction of factory automation (FA) technologies might be one way to solve these problems. However, the cost to introduce FA technologies is so huge that many working processes are still operated by human power. We should provide a method to prevent human error for processes when human power is dominant. Generally, human errors come from lack of worker concentration. This paper values body movement of workers while they show symptoms of human error. We think the worker have good rhythm and bad rhythm in monotonous work. The body movements of the worker detect rhythm. We have focused on the transition of the degree of similarity. The experiment makes workers to simulate repetitive works of tracing a circle shown on a tablet PC and displayed 100 times. This system finds a symptom of human error. We make sure that the transition of the degree of similarity has validity.

2 HUMAN ERROR IN MONOTONOUS WORK

2.1 *An assembly line method*

Work process has the problem of an accident which leads to injury of the workers and poor quality. In many cases, the cause is human error. Human error is the unintended fault of workers and it accounts for 80% of accidents. Therefore the goal is the eradication of human error. The tendency is particularly strong when the worker has a good rhythm, and they can repeat the same motion because their movements are refined assembly line method.

An assembly line method can produce a single product in large quantities in the factories. In this way, a worker repeats simple tasks such as nailing a screw. We can see its good examples in factories for assembly of automobiles and home appliances. The worker tends to make human error, because of reduced labor motivation from the meanness of the line production system.

There is a part that cannot be mechanized from the point of view of cost [1]. This paper focuses on human error in monotonous work. To prevent human error, plant manager works out "point and call or attention". However, it is inevitable to make human errors, thus it is necessary to detect the symptom of human errors. The purpose of this paper is directing human state just prior to human error. Validity is preventing human error to focus on worker's condition [2]. If the system finds a sign of error, plant manager can easily take action before an accident occurs.

2.2 *Rhythm of workers from body movement*

Even senior workers might make errors. Because of long monotonous tasks, they get tired and apt to lose their concentration. We assume a worker has his unique rhythm, when he engages in monotonous works. The disturbance of the unique rhythm has relationships with human errors. In the process of working, workers seem to take good rhythms, where they can work well without any mistake. After that, he disturbs the condition, because of various factors such as fatigue and loss of concentration. The red zone in Fig. 1 is a symptom of a human error. This research aims at finding the symptom of human errors easily. We propose a method analyzing the body movement. Since working quality and body movement are correlated, concentration is measured from body movement. Workers take certain movement in a good working condition. However, when they lose their rhythm, they will not repeat the same movement as they were in a good working condition. For example, they take unnecessarily

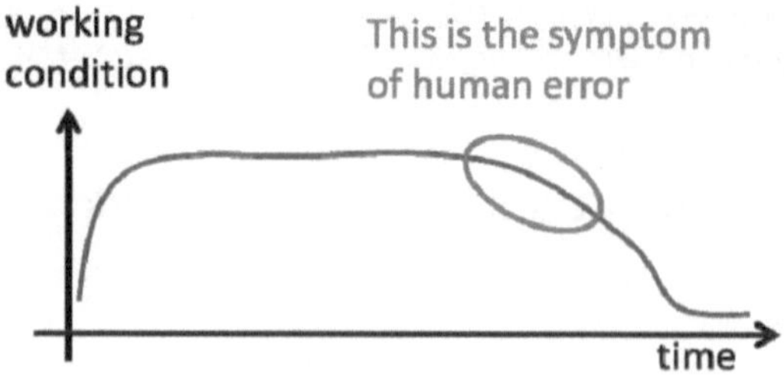

Figure 1. Worker condition.

large body movement, or face working sites from different angles. Disturbance of the body movement is a symptom of human errors. Analysis of body movement contributes to the detection of the symptom of human errors.

2.3 *Related works*

There are several researches which study the movement of the body of the worker to prevent human error. [3] proposes a technique to watch a worker using a camera. The camera detects worker movement to tighten a screw. The analysis of differences in the movement discovers a mistake. However, a camera gives a discomfort to workers. Moreover, a camera cannot shoot outside the scope.

In the method presented in [4], an acceleration sensor is attached to the chair. This acceleration sensor acquires action taking and detaching of workers. The method judges concentration from data acquired. However, a chair is necessary for this method. It cannot detect concentration of workers standing in working places. We need a wearable technique which can find concentration from any worker.

3 DETECTING HUMAN ERROR SYMPTOM FROM OPERATOR MOVEMENT

3.1 *Accelometor*

We propose a method to discover a sign of human errors from the movement of the body. Fig. 2 shows an overview of the method.

Workers wear the accelometer. The accelometer continuously acquire body movement of workers. The body movement is represented with 3 dimensional digital data, acceleration (x, y, z) and angular velocity (x, y, z), where x, y, and z represent the dimensions. When the body movement of the worker is disturbed, the system determines the symptom of human error. There is a research discovering abnormal movement using an accelometer that can detect a symptom of a human error. The accelerometer has three advantages:

- Low cost
- Specializing
- Wearable

It is feasible to attach an accelometer to a worker, because it is so inexpensive that it is equipped with a Smartphone. The body movement the accelometer

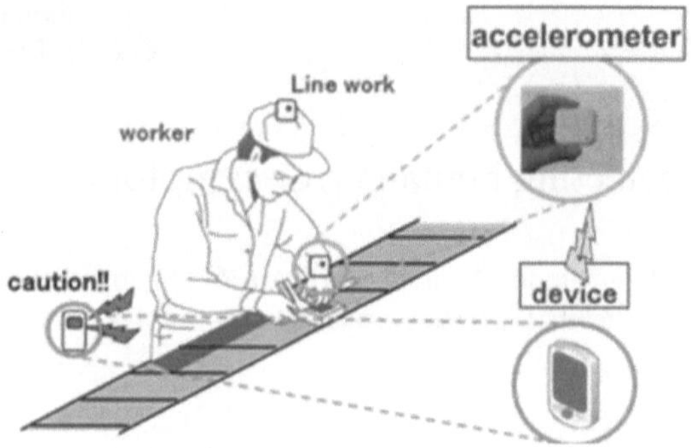

Figure 2. Overview.

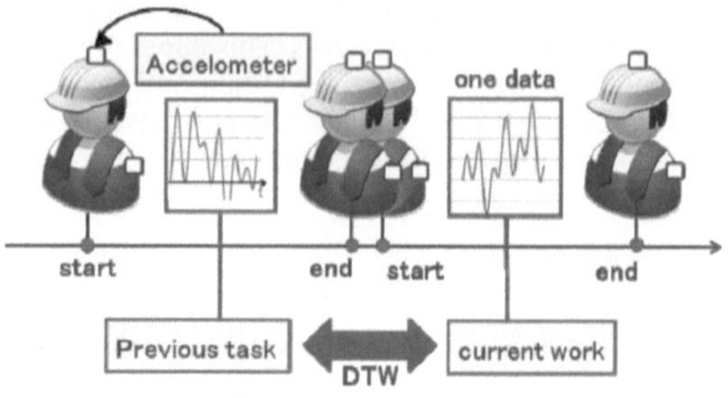

Figure 3. Compare method.

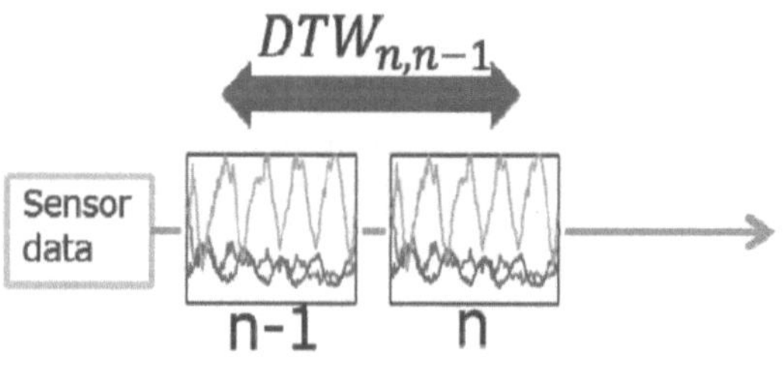

Figure 4. Define DTW.

acquires is independent from each worker. The symptom of human errors varies from person to person. Reflecting personal characteristics, we can enhance the precision to detect symptom specific to every worker.

Wearableness is also an advantage [5]. Wearable sensors can take characteristics of every worker. It improves the precision to detect personal symptoms.

3.2 *Working rhythm*

The monotonous work is a process to repeat short task. We focus on the transition of body movement in every short work. Regarding the short work as one period, the proposed method acquires a series of data representing the body movement of the worker during one period with several accelerometers. The method compares the series data of the current task with that of the previous task as shown in Fig. 3. The method calculates difference of the series data during the current task from that of the previous task. The smaller the difference, the better the working condition. The method also calculates the difference with the Dynamic Time Warping

(DTW). DTW is the Levenshtein Distance of a wave data during one period from that of another period. If DTW value is big, the current series data is quite different from the previous one. Fig. 4 shows the method of DTW. Suppose data during the current task and the previous task are denoted by n and n − 1, respectively. In this paper, $DTW_{n,n-1}$ is defined to represent the similarity of series data acquired in task n with that acquired in task n − 1. This system watches the evolution of the similarity. The transition of $DTW_{n-1,n}$, $DTW_{n,n+1}$, $DTW_{n+1,n+2}$, ... indicates the working condition of the worker. When it stays small enough, the working condition is expected to be stable, because it means the body movement of the worker is almost similar in successive tasks. On the contrary, when it gets larger, the working condition seems to be getting worse. The paper adopts the sum of five successive DTW as the working rhythm. It reveals symptoms of human errors. A worker is assumed to be in a dangerous state if the working rhythm is getting large.

3.3 *Naïve Bayes*

This system collects Working Rhythm both in case the worker has a symptom of a human error and the case he does not, as it is shown in Fig. 5. Suppose the worker has symptom, in the red zone in Fig. 5. This method separates body movement with an error symptom and body movement without it. This method uses the naïve Bayes [6] which is a supervised learning algorithm. Naive Bayes has a response variable and predictor variables. In the learning phase, a response variable is given as teacher data, while obtained data are predictor variables. In this method, the Working Rhythm is the predictor variable, and binary data indicating whether a symptom of human error appear or not is response variable.

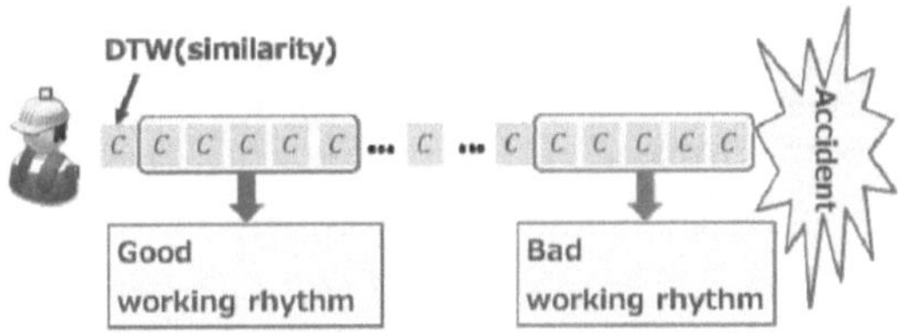

Figure 5. Tow working rhythms.

4 EXPERIMENT

4.1 *Environment*

To demonstrate the usefulness of the working rhythm, we have conducted an experiment regarding on a simple task anyone can accomplish. As a simple task in experiments, we adopt a task to repeatedly trace a circle drawn on the tablet PC. The task is easy enough to anyone. However, people out of concentration cannot draw a clean circle. Fig. 6 represents the screenshot of

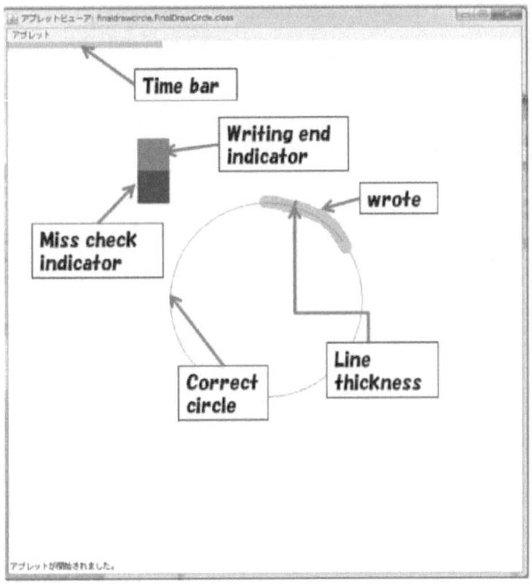

Figure 6. Display.

Table 1. Data acquired in the experiment.

symptom	x	y	z	ax	ay	az
non_error	946111	104064	969258	612524	2378984	342011
non_error	1364991	104839	1402482	579425	920592	335655
non_error	1361301	103519	1273259	576556	917299	355301
non_error	1360410	99223	1235570	519944	843087	336566
non_error	1312108	103458	1309804	601307	980968	310517
non_error	1896054	110658	1801015	582136	1152347	407409
non_error	1558659	125482	1348607	776853	1387351	674850
non_error	1577886	144036	1386819	1005198	1567449	807034
non_error	1737791	159622	1561560	1259176	1917599	896621
non_error	1705288	150457	1513743	1155976	1850515	888877
non_error	1576171	138816	1460120	1164902	1671046	888113
non_error	1932785	137061	1893552	1057584	1456728	677264
non_error	2438774	115708	2379559	768790	1473115	541759
non_error	2347071	104461	2223986	516631	1281825	466719
non_error	2309644	108151	2205383	501347	1209331	409987
error	2879503	104876	2639181	543930	1259994	280609
error	2406055	99302	2155356	570408	1245305	264120
error	3510444	111352	3321447	609104	1162569	283545

the tablet PC. The tablet PC displays the correct circle subjects should trace. Subjects trace a line of width w on the tablet PC with a stylus. In the experiments, w is set to 8 pixels. When subjects put the stylus on the tablet PC, the writing end indicator in the figure becomes red to show the tracing status. When subjects finish one circle, it turns into blue. There can be two kinds of failures in this experiment. One is the time over, while the other is to protrude a circle. In the experiment, subjects should finish the tracing within 7 second. Expression indicates the condition of the protruding error.

r is the radius of the correct circle. (x, y) is the coordinate of the center of this circle. (x_k, y_k) is the coordinate of the grounding point of the stylus. When the subject cannot meet this condition, the missing check indicator changes from blue to red.

Table 2. Example.

		actually	
X_point		error	non error
System	error	2	4
	non error	1	13

4.2 *Outline*

The Working Rhythm detects a symptom of a human error. In this experiment, we test the usefulness of the Working Rhythm. 13 subjects perform the simulated work explained in 4.1. In this research, the disturbance of the working condition comes from a long working time. In this analysis, we focus on subjects who have succeeded in more than 20 continuous tasks without any error. Subjects wear accelometers on the wrist and the head. Table 1 indicates data acquired in the experiment The first element (symptom) is the response

Table 3. Result A-F.

		actually				actually	
A_head		error	non error	D_head		error	non error
System	error	2	0	System	error	2	2
	non error	1	15		non error	1	50
		actually				actually	
A_hand		error	non error	D_hand		error	non error
System	error	1	0	System	error	1	0
	non error	2	15		non error	2	52
		actually				actually	
B_head		error	non error	E_head		error	non error
System	error	2	0	System	error	2	1
	non error	1	17		non error	1	17
		actually				actually	
B_hand		error	non error	E_hand		error	non error
System	error	3	16	System	error	2	1
	non error	0	1		non error	1	17
		actually				actually	
C_head		error	non error	F_head		error	non error
System	error	1	2	System	error	1	0
	non error	2	21		non error	2	14
		actually				actually	
C_hand		error	non error	F_hand		error	non error
System	error	2	3	System	error	2	3
	non error	1	20		non error	1	13

variable. Elements of the 2th to the 7th are the working rhythm values of the specified data items. These are predictor variable. We expect that the 6 predictor variable identifies the class of binary (error or non-error). This study uses the cross-validation, assuming one row in the table is the test data, while others are training data. It repeats in all rows.

4.3 *Result*

Table 2 indicates example. This method tries one point of subject X. A symptom of a human error appears three times. The table head indicates the actual events.

The table side indicates the system predication for the appearance of error symptoms. Table 2 indicates that the system could identify two of three symptoms of human errors. This system provides the correct prediction in 13 of 15 cases where any error symptom does not appear, as it is shown in Table 2. Table 3 and Table 4 show the results of the wrist and head of the 13 subjects from A to L. The system could detect a symptom of at least one error in all subjects. Let us investigate the sum of the transition of the similarity. The similarity is calculated with DTWs of the successive tasks. Let the number of successive tasks as the window size.

Table 5 compares the recall and the precision in case the window size is 5 and 1. When the window size is

Table 5. Validity of window 5.

error	recall	precision
window1	0.361	0.089
window5	0.528	0.485
non-error	**recall**	**precision**
window1	0.697	0.93
window5	0.936	0.956

Table 4. Result G-L.

		actually				actually	
G_head		error	non error	J_head		error	non error
System	error	1	5	System	error	1	0
	non error	1	33		non error	2	50
		actually				actually	
G_hand		error	non error	J_hand		error	non error
System	error	2	3	System	error	2	3
	non error	2	33		non error	1	47
		actually				actually	
H_head		error	non error	K_head		error	non error
System	error	2	3	System	error	2	0
	non error	1	76		non error	1	49
		actually				actually	
H_hand		error	non error	K_hand		error	non error
System	error	0	6	System	error	1	2
	non error	3	73		non error	2	47
		actually				actually	
I_head		error	non error	L_head		error	non error
System	error	2	0	System	error	2	0
	non error	1	13		non error	1	26
		actually				actually	
I_hand		error	non error	L_hand		error	non error
System	error	2	0	System	error	2	2
	non error	1	13		non error	1	24

equal to 1, nothing is summed up. This result indicates the precision and the recall of the total number of detections in all subjects. The case of window 5 is clearly superior to that of window 1. Therefore, the similarity calculated with successive task is effective.

5 CONCLUSION & DISCUSSION

We have suggested a method for detecting a sign of human error by the transition of the movement of the worker body in a monotonous work. The experiment has revealed that the working rhythm calculated from DTW in successive tasks is effective to predict human errors.

REFERENCES

Dept. of Comput. Sci. & Eng., Univ. of South Florida, Tampa, FL, USA: A Survey on Human Activity Recognition using Wearable Sensor, Communications Surveys & Tutorials, IEEE (Volume: 15, Issue: 3), 1192–1209, 2013.

Jeremy Rifkin: The End of Work: The Decline of the Global Labor Force and the Dawn of the Post-Market Era.

Masashi Okubo, Aya Fujimura. Development of Estimation System for Concentrate Situation Using Acceleration Sensor. Human-Computer Interaction. New Trends Lecture Notes in Computer Science Volume 5610, 2009, pp. 131–140.

Rios, Irina. "An empirical study of the naive Bayes classifier". IJCAI 2001 Workshop on Empirical Methods in Artificial Intelligence. 2001.

Shimizu, Sanae; Hirayu, Hidekazu; Asai, Hirotsugu. Development of Failsafe Camera System for Fastening Work in Manufacturing Scene IEEJ Transactions on Electronics, Information and Systems, Volume 129, Issue 5, 2009, pp. 916–922.

Zhiqiang Sun, Erling Gong, Zhengyi Li, Yingjie Jiang, Hongwei Xie: Bayesian estimator of human error probability based on human performance data, IEEE, Journal of Systems Engineering and Electronics, Vol. 24 No. 2, pp. 242–249, (2013).

Industrial Engineering and Management Science – Lee (Ed)
© 2015 Taylor & Francis Group, London, ISBN 978-1-138-02647-6

The design and evaluation of a mobile UI for personalized dietary food selection

Ying-Chieh Liu & Ya-Chi Tsou
Industrial Design Department, Chang Gung University, Tao-Yuan, Taiwan

Chien-Hung Chen
Institute for Information Industry, Taipei, Taiwan

Hsin-Yun Chen
Department of Nutriton Therapy, Chang Gung Memorial Hospital, Taipei, Taiwan

Yu-Sheng Lin
Internal Health Promotion Center, Chang Gung Memorial Hospital, Tao-Yuan, Taiwan

Cheng-Hung Lo
Industrial Design Department, Chang Gung University, Tao-Yuan, Taiwan

ABSTRACT: Two app interfaces are designed and evaluated for the input personalized dietary intake: a) a tree view structure with photos; and b) a list structure without photos. Trials with 30 university students indicate that both interfaces provide useful features. Future research will focus on design enhancements to improve the user experience for different target groups.

1 INTRODUCTION

Service industry companies are seeking to raise their competitiveness through mass customization creating products which can be customized based on a customer's specific needs (Salvendy, 2001). One example of this trend is the development of personalized dietary services in cafés or soft drinks shops, where the customers can customize their beverage based on their preferences for different flavors, ingredients and topping (Kamis, Koufaris, & Stern, 2008). Given increased awareness of and concern for healthy diet practices (Moustafa & Froguel, 2013), (Strobel, 2014), customers could find such services useful in helping them manage their weight and maintain a healthy lifestyle.

Dietary intake is an essential component of dietary management (Kuczmarski, Moshfegh, & Briefel, 1994), (Byrd Bredbenner & Schwartz, 2004), and dietary management apps have emerged as a very popular genre of downloads for smart mobile devices (Fox & Duggan, 2012), (Rusin, Arsand, & Hartvigsen, 2013), with users using the apps to plan, manage and track their dietary intake. However, despite the popularity of such apps, few studies have actually examined interface usability or usage patterns over time (Chesanow & Fogg 2013). Inappropriate interface design could result in app usage being time-consuming, inconvenient or even error-prone, thus reducing the likelihood that the app will actually be used on a sustained basis.

This study reviews two designs of innovative aids that allow users to build customized food items for dietary intake tracking. Test users were asked to interact with the different designs in tasks that emulated the input of individual meals so as to determine interface design factors which are conducive to effective and desired use.

2 DESIGN

An iterative user-centered design process is adopted from Bandura's social cognitive theory (1977) to investigate the impact of potential design interfaces on navigation efficiency under specific environmental conditions. The design approach entails proposing, evaluating, and selecting design alternatives so as to eventually obtain an effective and user-friendly design. The two designs were based on the stereotypical user model (Fischer, 2001) in which common user interfaces and interactions are defined based on existing apps to determine the basic building blocks for interface design. Potentials users were defined as university students and senior citizens capable of effectively operating smart phones or table computers. However, only university students were used in the evaluation process.

Most dietary aid interfaces fall into one of two types. In the first, category-based approach, users select from a list of pre-defined categories, and navigate through nested sub-categories to find the desired food item. In the keyword based approach, users type in keywords related to the desired food item to create a cluster of relevant selections which would potentially include the desired food item. This research did not consider the keyword based approach, and both tested designs were based on the category-based approach.

The two interfaces were designed with the following criteria in mind: 1) allow the user to find a selection of customized food items through nested categories, or 2) use food icons as visual cues to help users identify the appropriate food category or sub-category without the need for supplemental guides or memory aids.

Personalized foods items were classified into two categories, including beverages (customizable in terms of toppings, type of milk, and sugar content), and mixed foods (customizable in terms of combining two or more specific foods as a dish). Mixed foods are commonly found in Asian cuisine (e.g. Chinese cuisine).

The interface designs are also divided into two categories, tree and list, as follows:

2.1 *List structure for beverages*

Figure 1 illustrates the list structure without photos interface operation. In Figure 1a, the user browses up or down to click for selecting a specific food category (e.g. soft drinks category), followed by a sub-category (e.g. coffee sub-category) in Figure 1b. The user then selects an appropriate food item (e.g. coffee mocha) as shown in Figure 1c. The user is then asked to select from a list of customized food items that are the same food item with different combinatorial set of toppings (e.g. coffee mocha, little sugar, and low fat milk). The user then completes the activity and submits the data for caloric calculation (assuming a certain food portion) and storage as shown in Figure 1e. A highlight of the button area will be shown after user's touch on the list of the selection. Also, a return button allows the user to go back to the previous action.

2.2 *Tree view structure for beverages*

As shown in Figure 2a, in the tree view structure with photo interface, the user selects a category by choosing the appropriate icon. As shown in Figure 2b, the user then navigates the nested sub-menus, and selects an appropriate beverage item (e.g. coffee mocha) as shown in Figure 2c. Once the user clicks the item, a tabbed structure is presented to allow the user to customize it (i.e. through selecting toppings, milk and sugar content, respectively shown in Figs. 2d, 2e, and 2f).

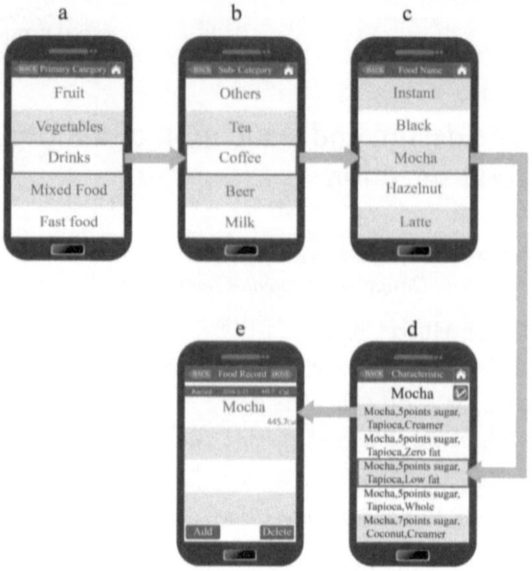

Figure 1. List structure without photos.

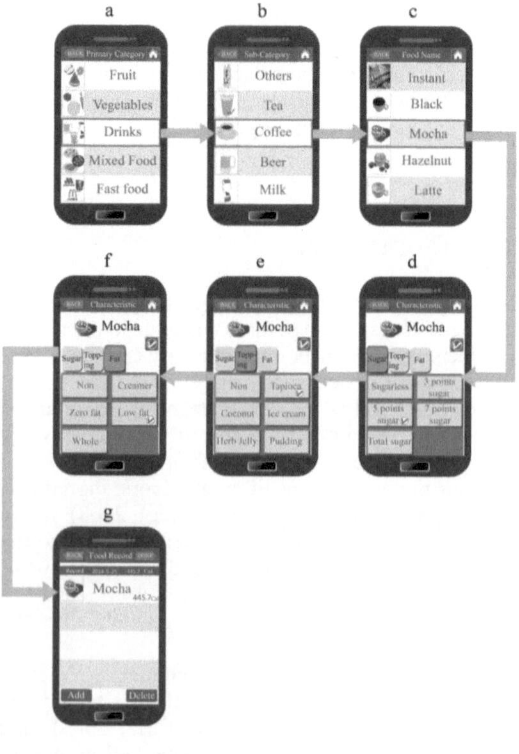

Figure 2. Tree view structure with photos.

2.3 *List structure for mixed foods*

For recording mixed food intake using the list structure, the user first selects the appropriate food categories (Fig. 3a) to produce a list of mixed food items with various preparation modes. The user browses

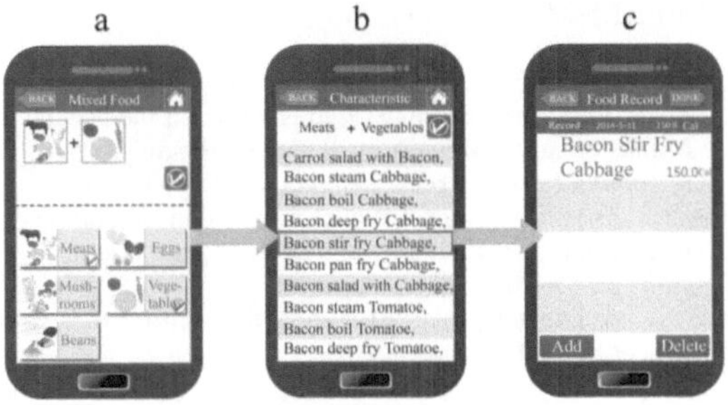

Figure 3. List structure without photos.

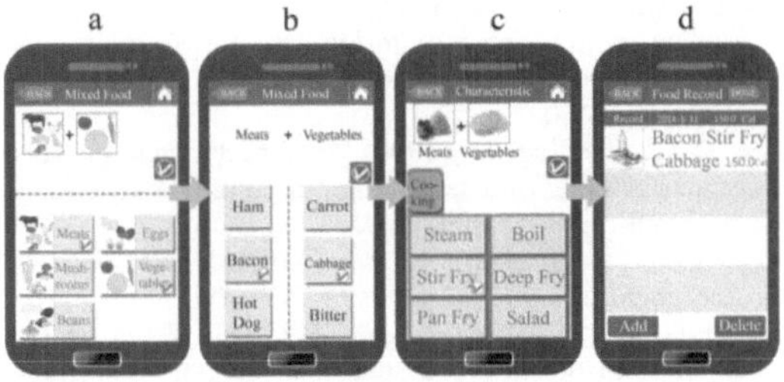

Figure 4. Mixed food intake recording process based on tree view structure with photos.

these choices and makes the appropriate selection (Fig. 3b), and confirms the selection (Fig. 3c).

2.4 *Tree view structure for mixed foods*

For mixed foods, the user first selects the Mixed Foods category to call up two empty fields. The user then clicks each field and selects from potential food categories (e.g. meat) shown in the lower section of the interface (Fig. 4a). Having determined two food categories, the user then browses two scrolling food lists and selects the desired foods (Fig. 4b). The user then selects the mode of preparation (Fig. 4c) and clicks confirm to complete the food selection process (Fig. 4d).

3 INITIAL SUBJECTIVE EVALUATION

An initial subjective evaluation using open-ended questions was conducted to assess interface preference. We recruited 30 students at Chang Gung University in Taiwan (9 males and 21 females) between 18 and 29 years old. The test group was split into two subgroups to test the two interfaces, with group A (12 females, 6 males) testing the tree-view structure with photos interface, and group B (9 females 3 males) testing the list structure without photos interface. All test subjects were asked to use the interface to search for and find a pre-determined selection of foods on a mobile device with a 7 inch touch panel. Subjects were then asked to complete a questionnaire with four open-ended questions to determine their interface preferences. All the subjects were presented

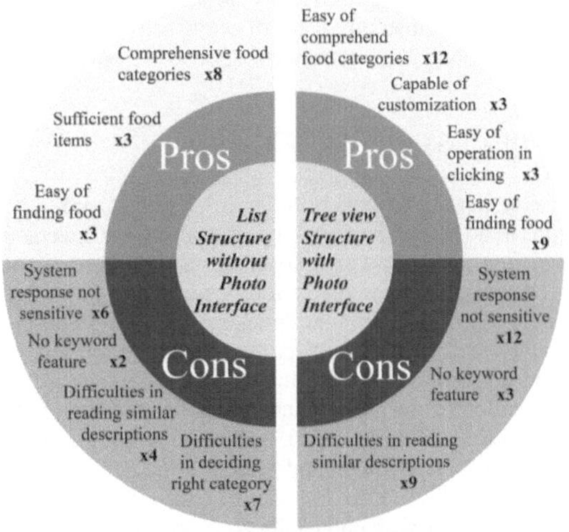

Figure 5. Thematic network for both interfaces.

with two meals using real food items. Each food item was equipped with a paper label displaying the name of the food item and its ingredients (e.g. coffee mocha, a little sugar, low fat milk). Each meal contained a total of six discrete food items, selected in consultation with a senior dietitian to represent a typical complete and nutritious meal for adults. Before the test, interface operation was explained to all subjects, providing time for them to try the interface and ask questions. To ensure the test subjects felt comfortable using the system prior to the test, each subject was then asked to try to input a pre-test meal containing six food items (different from those in the test meals). The 12 test food items were steamed bread, chicken, green tea, tofu egg salad, carrot fried eggs, bacon fried cabbage, rice, pork chops, milk tea, shredded green pepper, scrambled eggs with tomato, and bitter gourd fried salted duck eggs.

4 RESULTS

The thematic network method (Lee & Fielding, 1996), (Martin & Hanington, 2012), was used to summarize subject responses to both interfaces (Fig. 5): tree view structure with photos and list structure without photos. Advantages cited for the list structure with photos included comprehensive food categories (8 subjects), sufficient food items (3 subjects) and ease of finding foods for recording (3 subjects). Subjects also indicated concerns regarding difficulty in determining the correct food category from the category list (7 subjects), interface response times need to be reduced to improve browsing (6 subjects) and difficulty finding similar food descriptions (4 subjects). As for the structure view with photos, advantages included support for identifying food categories (12 subjects), ease of finding foods for recording (9 subjects), and ease of type-in operation (3 subjects). Some subjects reported concerns including difficulty identifying food categories (9 subjects), lack of keyword features (3 subjects), and slow interface response (12 subjects).

5 DISCUSSIONS & CONCLUSIONS

The respective groups noted similar design pros and cons in both interfaces, but the structure view interface was found to have a significant advantage in terms of presenting comprehensible food categories, likely due to the use of food images as visual cues. One notable drawback for the list structure is that the similar of food descriptions is hard to read, and the severity of this problem would be exacerbated by the inclusion of more food items. Also, subjects in both groups cited difficulty finding and selecting the desired food categories despite the inclusion of visual cues in the structure view with photos interface. A possible explanation is that the users' limited background knowledge makes it difficult for them to find sufficient information from visual cues or they are simply not aware of the food items represented by the images. Such problems could potentially be resolved through other strategies, e.g., providing a keyword feature. Also, interface responsiveness needs to be improved.

Issues related to task complexity could be explored by expanding the variety of food items in the menus, and both interfaces require further usability improvements and testing with different user groups to objectively evaluate selection errors and operating time requirements.

ACKNOWLEDGEMENTS

The authors wish to thank student-testers. This research is partly funded by the Chang Gung Memorial Hospital (CMRPD2C0021) and the National Science Council of the Republic of China (NSC 102-2410-H-182-014).

REFERENCES

Byrd Bredbenner, C., & Schwartz, J. (2004). The effect of practical portion size measurement aids on the accuracy of portion size estimates made by young adults. *Journal of Human Nutrition and Dietetics, 17*(4), 351–357.

Fischer, G. (2001). User modeling in human–computer interaction. *User modeling and user-adapted interaction, 11*(1–2), 65–86.

Fox, S., & Duggan, M. (2012). Mobile health 2012. *Pew Research Center's Internet.*

Kamis, A., Koufaris, M., & Stern, T. (2008). Using an attribute-based decision support system for user-customized products online: an experimental investigation. *Mis Quarterly*, 159–177.

Kuczmarski, M. F., Moshfegh, A., & Briefel, R. (1994). Update on nutrition monitoring activities in the United States. *Journal of the American Dietetic Association, 94*(7), 753–760.

Lee, R., & Fielding, N. (1996). Qualitative data analysis: representations of a technology: a comment on Coffey, Holbrook and Atkinson.

Martin, B., & Hanington, B. (2012). *United States of America: Rockport* (Universal Methods of Design ed.).

Moustafa, J. S. E. S., & Froguel, P. (2013). From obesity genetics to the future of personalized obesity therapy. *Nature Reviews Endocrinology, 9*(7), 402–413.

Rusin, M., Arsand, E., & Hartvigsen, G. (2013). Functionalities and input methods for recording food intake: A systematic review. *International journal of medical informatics, 82*(8), 653–664.

Salvendy, G. (2001). *Handbook of industrial engineering: technology and operations management*: John Wiley & Sons.

Strobel, M. (2014). The Future of the Weight Management Industry Retrieved Jan 2014, from http://blog.euromonitor.com/2014/01/the-future-of-weight-management.html.

Industrial Engineering and Management Science – Lee (Ed)
© 2015 Taylor & Francis Group, London, ISBN 978-1-138-02647-6

The credit risk assessment of P2P lending based on BP neural network

DunGang Zang
College of Economics & Management, Sichuan Agricultural University, Chengdu, China

MngYu Qi
College of Software, University of Science and Technology of China, Hefei, China

YanMing Fu
School of Computer, Electronics and Information, Guangxi University, Nanning, China

ABSTRACT: P2P lending is a new type of financial model under current Internet financial circumstance. The evaluation of P2P Internet loan has tremendous meaning since its credit risk is the leading factor of Internet finance stability. According to the characteristics of P2P loan, BP neural network model was introduced to evaluate the risk of credits quantitatively. We build an empirical model with the lending club data. Given the experimental result, P2P Internet loan credit risks are primarily determined by several key attributes. In addition, BP neural network method offers an evaluation result with 78.6% accuracy rate (type one falsification rate 4.8% while type two falsification rate 16.6% respectively), which indicates that risk evaluation method has a predominant performance.

Keywords: P2P lending; credit risk assessment; BP neural network; principal component analysis.

1 INTRODUCTION

P2P lending refers to small unsecured loans between individual and individual, which establishes lending relation and completes the related formalities directly through the internet platform without banks and other financial institutions as intermediaries[1]. P2P lending is different from traditional lending. P2P lending has the characteristics of low threshold of loans, wide coverage, fast circulation of information, convenient trading formalities, small amounts involved, short loans period and so on. These incomparable advantages over formal financial institutions solve the financing problem of small and medium-sized enterprises, which is called online edition of folk lending. However, network virtual characteristic, information asymmetry and imperfection of risk control measures platform increase the possibility of borrowers defaults, because P2P lending is based on network platform, which increases the credit risk of P2P lending. Therefore evaluating the credit risks of the P2P lending has important practical significance for maintaining the stability of financial system.

From the existing literature available, the research on P2P credit risk mainly focused on qualitative analysis, lacking researching on quantitative evaluation. Foreign scholars, named R Vedala, classified naive Bayesian, and used them on financial statements data indicators in 2013, establishing credit risk assessment model of P2P. This method involves how to train the accurate probability problem[2]. Domestic scholars, named ChunYu Luo, proposed the risk assessment model based on kernel, we can get the result of loan risk and yield rate by using after training regression model based on kernel function, this model has a higher prediction accuracy relative to linear regression model frequently used in traditional lending[3].

Integrating current literature about credit risk assessment research of P2P lending, we can see that few scholars use the BP neural network to evaluate the credit risks of the P2P lending. This paper establishes an effective credit risk assessment model by using BP neural network, based on characteristics of P2P Business model, which provides enlightenment for risk control mechanism of P2P lending platform and provides support to solve the problem about the risk control of P2P lending platform construction.

2 BP NEURAL NETWORK THEORY

BP (Back Propagation) Network is proposed by a group of scientists headed by Rumelhart and McCelland in 1986. BP Network is a kind of multilayer feedforward network based on deviation inverse propagation algorithm training, which is one of the most widely used neural network model. Topological structures of BP neural network model include input layer, hide layer and output layer, as shown in Figure 1[4].

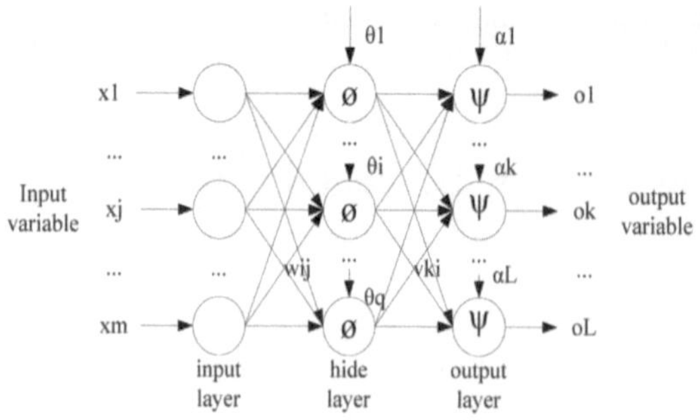

Figure 1. Neural network model.

Figure 1: xj is denoted as the input of the first j node of input layer, j = 1, . . ., M; w_{ij} is denoted as the weight between the first i node of hidden layer and the first j node of input layer; θ_i is denoted as the threshold of the first i node of hidden layer; $\phi(x)$ is denoted as the excitation function of hidden layer; v_{ij} is denoted as the weight from the first k node of output layer to the first i node of hidden layer, i = 1, . . . , q; a_k is denoted as the threshold of the first k node of output layer, k = 1, . . . , L; $\psi(x)$ is denoted as the excitation function of output layer; O_k is denoted as the output of the first j node of output layer.

BP neural network is a supervised learning algorithm, the main idea of it is inputting learning samples, repeatedly adjusting training the weights and deviation of network by using back propagation algorithm, to make the output vector and expect vector as much approximative as possible. When the deviation square of output layer is smaller than the specified deviation, the training has been completed, saving the network weights and deviation finally. The realization of the BP Neural Network includes the following two steps.

(1) The procedure of forward propagation of signals
Enter variable x into the network through input layer neurons, get the output θ according to weight w and excitation function of hidden layer neurons Φ, and make the output θ as the input of output layer neurons. Get the output o according to weight v and excitation function of output layer.
(2) The procedure of deviation back propagation
Back propagation of deviation: first calculate output deviation of each layer neurons step by step from output layer, then adjust the weight and threshold of each layer according to deviation gradient descent method to make the revised final output of the network close to the expectation.

3 MODELING OF CREDIT RISK ASSESSMENT

BP neural network is a classification problem, loan record can be divided into normal repayment and abnormal repayment. Determine the structure of BP neural network in view of the practical problems, then train the BP neural network, thus we can get a credit risk assessment model with higher detection accuracy.

Table 1. The distribution of data sets.

	Scale of data set	Normal payment record	Abnormal payment record
Training data set	6669	4793	1876
Test data set	3800	3000	800
Overall data set	10649	7793	2676

The attribute of each loan records in data set transform Dimensionless value after pre-processing. At this time, the remaining properties in addition to the classification attributes constitute an input vector. Namely $X = (x_1, x_2, \ldots, x_n)$, n is the number of attributes. X is denoted as neurons of input layer, the output of the BP neural network is a two-dimensional vector: (1, 0) represents normal repayment, (0, 1) represents abnormal repayment.

4 ANALYSIS OF MODEL EXPERIMENT

The data used in this paper is a collection of loan records downloaded in P2P lending website (lending club1) in United States, which is the latest data between January 1, 2012 and April 30, 2014.

4.1 *The composition of data set*

According to the value of the status attribute loan_status, pick up all the due record of loans to constitute a empirical data set, timely repaying records in all expired records are called as normal repayment record, otherwise are called as abnormal repayment record. Choose records randomly from the data collection to form respectively training collection and test collection of the model. The data distribution of Test set and training set as shown in Table 1.

4.2 *Pretreatment of data sets*

Considering that the loan records of the lending club has more than 100 attributes, choose the attributes that have great influence on repayment status as input attribute of the model. As shown in Table 1, we get the result through Principal component analysis, which indicates that the model built by seven attributes can make evaluation precision converge to a model built by attribute set that has an obvious influence on repayment status. As shown in the following, seven attributes are chosen from the record according to the result of the experiment: loan_amnt (the loan amount of application), term (allotted time of the loan), int_rate (loan interest rates), grade (level of credit), emp_length (the time engaged in this work), home_ownership (have a house or not), and annual_inc (annual income), and the addition of loan_status (status of loan), which is a classification attribute.

Convert Seven attributes to be numeric. Loans can be divided into 36 months and 60 months, expressed in numerical 3 and 5 respectively. Level of credit can be divided into five types from high to low: A, B, C, D and E, and the relative numerical value are 1, 2, 3, 4 and 5. Working time of every year can be regards as a level, this paper regards it as 11 when the working time (emp_length) is more than 10 years. For the ownership of the house (home_ownership), renting is given weight 1, owned housing is given weight 5. For loans status, the status that repays the loan on time is regarded as 0, other statuses are regarded as 1, which represents deferred repayment. The attribute of loan amount, loan interest rates and Annual income are numeric data for themselves, without conversion.

In matlab, use the function Y = zscore(X) for data standardization, to eliminate the dimensional difference between the data.

4.3 *Procedure of experiment*

In order to import the data sets of table 3 and implement training and testing of BP neural network, in this paper, the experiment uses neural network pattern recognition toolbox of matlab. The BP neural network with single hidden layer structure has an optimal prediction effect. The output is a binary classification attribute, (0, 1) represents abnormal loan, while (1, 0) represents normal loan. Therefore, output layer needs to use two neurons. Determine the number of hidden layer node according to the following empirical formula.

$$n_1 = \sqrt{n+m} + a \tag{1}$$

In the formula, n represents the number of input node, while m represents the number of output node, a represents a constant between 1 to 10. In this paper, determine ten hidden layer nodes according to experience. Use sigmoid function as Excitation function.

$$f(x) = \frac{1}{1+e^{-x}} \tag{2}$$

This paper uses principal component analysis[5] to analyzes the data set, and then test the relationships between the number of attributes and detection rate. In matlab, use the function princomp to implement principal component analysis. Principal component analysis[5] is a statistical method of dimension reduction. This method can convert the original random vector of component related to a new random vector of component unrelated by using orthogonal transformation method, which can transform it into a low-dimensional variables system. Then transform low-dimensional system into single-dimensional system further by constructing a proper value function.

We can learn from figure 2 that detection accuracy gradually increases when we gradually increase the number of attributes. But when the number of attribute

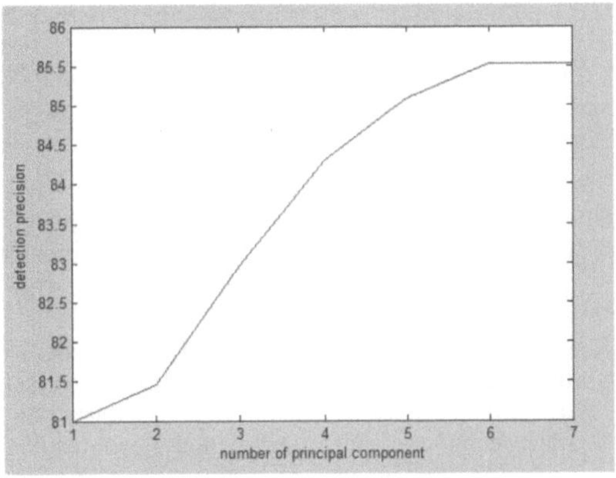

Figure 2. The relationship between the number of principal components selected and detection accuracy.

Table 2. Test results comparison of the model.

Accuracy	The first kind of error rate	The second kind of error rate
78.6%	4.8%	16.6%

is added to seven, the detection accuracy tends to convergence, added attributes cannot obviously improve the precision any more. As attribute increases, the running speed of BP neural network model also slows down, Seven attributes are the best input attribute combination with a comprehensive consideration of accuracy and speed. Thus we form a three-layer BP neural network model of 7-10-2 type.Through the above experiment, the detection accuracy of the neural network model are determined as 78.6%.

4.4 *Analysis of experimental results*

Based on the data of Table 1, apply the BP neural network to assess credit risk, collect the accuracy, the first kind of error rate, the second type of error rate, and make a comparison among the data of these three indicators. Among them, the first kind of error rate represents that risky loan records are identified as loan records of no risk, while the second type of error is the opposite. Test results are as shown in Table 2.

BP Neural Network can obtain a higher recognition rate of credit risk, which has a better control on the second kind of error.

5 CONCLUSION

This paper demonstrates that BP Neural Network can overcome a series of problems appearing in traditional statistical method, such as lots of assumptions, the difficulty in counting parameter of measurement and so on, which demonstrates BP Neural Network is more suitable for exploration of nonlinear regularity. BP Neural Network is just suitable for the assessment of

P2P lending credit risk because its good predicting accuracy, low requirements for sample data and so on.

Based on the above relevant conclusions of P2P credit risk assessment, the following strategies can be adopted to improve the current situation of P2P credit risk assessment.

(1) P2P credit risk assessment of the neural network can only identify that whether there are credit risks in a loan record at a certain accuracy, so the evaluation result should not be directly regarded as a conclusion. But it can provide important quantitative indicators, The mediation of P2P lending can combine customer's personal information with macroeconomic indicators to make an accurate and reasonable judgement on customer credit risk.

(2) There are different attributes that describe loan records of P2P client, principal component analysis reveal that P2P loans credit risk is mainly decided by one of the few key attributes. Therefore, we can only take some key factors of them into consideration when analyzing the credit risk, which can eliminate the interference caused by secondary attributes to make a decision and maintain a P2P accuracy of credit risk assessment at the same time.

REFERENCES

[1] Wang Hui, Greiner Martina, Aronson Jay E. People-to-People Lending: The Emerging E-Commerce Transformation of a Financial Market [C]. 15th Americas Conference on Information Systems, 2009(6).

[2] Luo Chun Yu. Investment Decision Models in P2P Lending [D]. Doctoral Dissertation of Dalian University of Technology, 2012.

[3] Vedala, R. An application of Naive Bayes classification for credit scoring in e-lending platform [C]. Proceedings 2012 International Conference on Data Science & Engineering, 2012.

[4] Satish Kumar. Neural networks [M], McGraw-Hill, 2006 (3).

[5] Huang Qin, Chang Wei, Liu Yi-liang, Liu Yan-peng. BP Neural Network Classifier Based on Principal Component Analysis [J]. Journal of Chongqing Institute of Technology, 2009(7).

[6] Shao, Haihong; Ju, Xiaofeng; Li, Yukun. Study on Credit Risk Assessment Model of Commercial Banks Based on BP Neural Network [C]. International Conference on Frontiers in Computer Education, 2011.

Industrial Engineering and Management Science – Lee (Ed)
© 2015 Taylor & Francis Group, London, ISBN 978-1-138-02647-6

Introduced trade terms in the domestic trade to promote further development of the integration

Jiang-Nan Zhai & Wei-Biao Zeng
Hunan Radio & Television University, Changsha, China

ABSTRACT: The integration of foreign and domestic trade is a basic national policy of China as well as a commitment towards WTO. The integration process is slow though it began ten years ago. The main problem is that it's hard to unify the operation mode of the enterprises dealing in foreign trade and domestic trade, to dock the trade and to transfer the market. We should take the adoption, digestion and transplant of the trade terms as a breakthrough, and exert price mechanism to mobilize the initiative of the enterprises dealing in domestic trade, and create opportunity for those dealing in foreign trade to exploit domestic market.

1 INTRODUCTION

China's reform and opening up has entered deeper waters, which is now in overcome difficult stages. In the structural transformation of economy, there are lots of problems in circulation field. It's necessary to continue to emancipate the mind further. We should take the advantage of the international business achievements, import, digest and transplant new technologies, get rid of the path dependence formed in a segmented situation between foreign and domestic trade in the old system, so as to establish new trade ways in conformity to the international mainstream.

2 MACRO-ENVIRONMENT ANALYSIS

Ten years ago, at the beginning of the new century, the contradiction between the old superstructure and the new economic foundation was sharp and a reform with Chinese characteristics took place in China. In 2003, the integration of foreign and domestic trade was considered as a national policy in accordance with the rules of market economy and WTO requirements. With the effort of experts and scholars, the implication of integration of foreign and domestic trade is clear. "It's a process of economic development taking the enterprise as subject, the scale of market supply and demand and industrial division as a basis, the integration of the production elements market and product market at home and abroad as basic components, the market competition as the major driving force, and inducing the government management system and management policy to change collaboratively"[1].

It's a long road for the integration of the foreign and domestic trade, and now it's just a prelude[2]. However, the pace of integration has not stopped. Meanwhile the researchers in the industry have been constantly looking for energy to activate the integration. After years of research, people focus on "trade terms". As an "activation factor" of the integration, trade terms play a facilitating role though they only explore issues of cargo trade. Its application in domestic trade can drive the development of whole industrial chain including modern service trade.

With the deepening of the reform and opening up, more and more policies about circulation are published. On September 1, 2012, the State Council issued "the Twelfth Five-year Plan of Domestic Trade Development", and set eight main tasks in the circulation reform, which are specific measures to rationalize the domestic trade. In 2013, the government proposed to "promote the opening-up policy in foreign and domestic trade, the combination of import and export; boost the free flow of factors at home and abroad, effective allocation of resources and deep integration of markets."

From the perspective of the economic development pattern, the circulation industry, as a leading and basic industry of the national economy, plays a more and more important role in expanding domestic demand, promoting consumption, and guiding production, increasing employment, booming the development of urban and rural areas and perfecting the market. The integration of foreign and domestic trade in circulation industry is highlighted than ever before.

3 ENTERPRISES DEMANDS ANALYSIS

3.1 *The enterprises dealing in foreign trade have inner demand to exploit the domestic market under the squeeze of international market*

The production scale of the export-oriented processing trade decreases progressively due to the aftermath of the international financial crisis, the international trade protectionism and the fact that China has become the second largest economic entity in the world, which challenges the development and even living space of some export-oriented enterprises. In the meanwhile, with the improvement of living standards, more and more consumers buy the goods imported from China in overseas, which offers chance for the export-oriented goods in domestic market. The enterprises dealing in foreign trade face lots of barriers to exploit domestic market. The transaction mode and even the trade language are somewhat different. The physical goods, which can be transacted with the transfer of the bill in foreign trade, can only be traded with the movement of the goods itself and sometimes the deal cannot be made. The trade terms prevailing in foreign trade vanish in domestic trade[3].

3.2 *The enterprises dealing in domestic trade have internal impetus and external opportunity to promote market competitiveness under the promotion of structural transformation of the economy*

In the past, we went after the former Soviet Union and Eastern European countries, separating production and distribution, dividing the circulation industry into commercial areas of food supply and marketing cooperatives, food department, materials department and so on, which is a typical product of planned economy and never appears in the United States, Japan, Western Europe and other developed countries[4].

Compared with those of developed countries, the major economic indicators of China's circulation industry have been in an overall state of backwardness, with which are featured slow turnover rate, low circulation efficiency and high management costs[5,6]. They have a low rate of contribution to the national economy. The classical trade theory thinks that domestic trade is the foundation of foreign trade and foreign trade is an extension of domestic trade. Bad domestic trade level will restrict foreign trade competitiveness. As a domestic enterprise, the integration requires not only to exploit the international market but also to learn and master a set of effective methods and measures applied in international trade, such as international trade practices. To some degree, the application of international trade practices in domestic trade is the key factor for the integration of the enterprises dealing in domestic trade.

4 THE BREAKTHROUGH OF THE INTEGRATION

As previously mentioned, under division of domestic trade and foreign trade in circulation industry in the long term, each of them forms a set of path and method in the market, and there're needs for fusion with each other's and also learn from each other, but there is always a piece of invisible "glass ceiling" between domestic trade enterprises and foreign trade enterprises. A range of research results have been published in order to demonstrate the necessity to break "glass ceiling".

During the gestation period and initial period of the integration of foreign and domestic trade, theory researchers have repeatedly demonstrated from "Internalization of International Trade"[7] to Internationalization of Domestic Trade"[8], and solved the direction of the integration.

In order to find the right way in practice, some academics recommended "to use trade terms in domestic trade"[9]. For example, "On the Application of International Trade Terms in Domestic Trade Based on a Case,"[10] "On the Domestication of EXW",[11] "On the Application of EXW in domestic transaction"[12].

Some entrepreneurs defined new trade terms according to the INCOTERMS so as to regulate the transaction behavior[13], but the voice is weak. Compared with the current INCOTERMS, the new trade terms lack unity, quasi-mandatory and universality and is bound to be a flash in the pan.

Therefore, it's the highlight to combine the external pressure and internal need of exploiting domestic market of the enterprises dealing in foreign trade and the internal impetus and external conditions of boosting the market competitiveness of the enterprises dealing in domestic trade, which can be accomplished with the promotion of INCOTERMS 2010 rules in the enterprises dealing domestic trade.

4.1 *The INCOTERMS 2010 is a powerful tool to unify domestic and foreign trade operations*

The INCOTERMS 2010 rules are one of the international trade practices, which can be powerful tools for narrowing the distance standards disparity between foreign trade and domestic trade. Due to its quasi-mandatory, trade terms can be used flexible according to actually needs change. In this way, trade terms can become not only a connecter combining theory with practice but also filler for "breakpoint", and it would be used as a tool to pry off the "glass ceiling" between domestic trade and foreign trade in circulation industry.

Trade terms exist in system. Trade terms said above are referred to in existing systems together, rather is part of the "INCOTERMS 2010". Like all international practices, trade terms in "INCOTERMS 2010" are one of the sources of international law, which itself is not legally mandatory. Both sides according to the actual situation in interspersed with, after the agreement, confirmed by contract, only to have mandatory. As such, it builds a buffer of a moral line before legal line and provides great flexibility for trade agreement that will not be easily conflict with the bottom line, so as to improve the business efficiency.

Precisely because the trade terms have a natural of quasi-mandatory and its impact on the existing legal system is not as large as expected, it is not necessary to desolate trade terms just because their identity of "imported goods", or to neglect the distinctive effects using system of trade terms or to address of inefficient circulation transactions vice versa. Instead, one of the most important reasons for international practice to exist and develop is that it has been recognized by law in most countries, or compatible with the basic principles of law. Actually, using another a "imported goods"–"Learning By Doing" theory, Government and scholars has responsibility to encourage and guide enterprises which as market subject for marketing mode innovation in operating, under a condition of both traders to be familiar with trade terms, enterprises carry out trade terms boldly in domestic trading, learning by doing, tip by tip, a dominant results must will appear.

4.2 *The popularization of the application of trade terms in the enterprises in our country dealing in domestic trade is only a matter of time*

The formation and development of the trade terms experiences a long cultivation, historical accumulation and technology market tempering. Trade terms are the results of market economy, in line with the WTO rules; it is a foundation to improve the efficiency of goods circulation, to share resources of social, to form within the balance of domestic trade and foreign trade enterprise linkage development, to realize domestic trade and foreign trade patterns of the two-wheel-drive economy.

From an opinion of macro perspective, overall progress in the integration of foreign and domestic trade in circulation industry had reached a consensus. From an opinion of middle perspective, it is also easy to reach a consensus to take the domestic enterprises as the main aspects of the conflict. From a microscopic point of view, it is a new formulation that domestic enterprises use trade terms for integration as a breakthrough, or it is a benevolent thing. But this is in keeping with basic spirit "with plenty of guts, with steady of pace, combined top-level design with touching the stones for pass river, overall progress and breakthroughs in the promotion each other, make decision more scientific, broad consensus and form a reform force".

Trade terms are the result of market economy, first appeared in the United Kingdom in domestic trade, from birth up to now over 200 years of history[14]. In the system of the world's three largest trading terms, only The General Rules developed and lead by the International Chamber of Commerce is vital. The General Rules for change after more than 80 years since the birth, it not only draws the highlight of the "1932 Warsaw-rules of Oxford" and "the 1941 United States foreign trade", but also it has been revised with social and economic development and is recognized and adopted by more and more countries, forming international business development the social mainstream[15]. Everyone no matter what both sides, no matter what country, no matter under what legal system is available through the trade terms and reduce transaction costs to get win-win or more wins.

At the important juncture of our economic restructuring, as a responsible big country, it needs to change an opinion of "Heavy outside and light within", to develop market and product through that turning the threat of "population bonus" into superiority in increases ability of "skilled workforce bonus" and that using the characteristic of unbalanced development of coastal, Central and Western regions.

In this intensive growth all enterprises needs trade terms no matter both export-oriented enterprises in coastal areas who uphold the principle of "Internalization of International Trade" strategy or enterprises of the Midwest regional who exercise the strategy of "Internationalization of Domestic Trade" to carry task on gradient transference on the processing trade from coastal areas.

Moreover, the function in trade terms not only play a bridging role in integration of foreign and domestic trade, it can also promote enterprise to realize "Pareto improvement", practice skills, and strengthen fine and intensive management which can get internal benefits. When the Pareto optimal reaches within is the date that realizes integration of foreign and domestic trade in circulation integration.

In the accelerated process of global economic integration, it is very importance and self-evident that domestic and foreign trade enterprises in China adhere to the same rules[16]. It is well known that WTO rules were made for the gradual elimination of trade barriers. The trade barriers are not only including tariff barriers and Non-tariff barriers by artificial, but also including those the formation trade barriers because people not familiar with the rules of trade.

It's the basic mission of the integration to achieve the eight missions set in the Twelfth Five-year Plan and to implement the INCOTERMS 2010 rules in domestic trade. The trade terms can help to promote the circulation industry into a brand new field. Therefore, it is optimistic that the universal application of INCOTERMS 2010 rules in the enterprises in our country is only a matter of time.

5 SEVERAL CONSIDERATIONS

5.1 *The trade terms are price terms and should be applied with the combination of price strategy in marketing theory*

There're two mainlines in China's existing domestic goods pricing system. One is the price strategy in marketing theory, which is dominant and directs the enterprises to set the product price based on the following three orientations, namely, cost orientation, competition orientation and demand orientation. The other is the trade terms in international trade theory, which is recessive and considered as international conventions and directs the enterprises to set the product price in international trade[17,18].

The two pricing systems are like two parallels since they were introduced into our country. Comparatively speaking, the pricing system in marketing theory is dominant and has fundamentality, simplicity, unipolarity, single-sale certainty, stiffness, preference and home trade universality.

The fundamentality is the basis of pricing strategy and sets basic scope of the product price. The bottom line is the cost price; the top price is the demand price and the competition price is based on the market. The simplicity is an inherent ill of this system. It has very limited connotation and only considers price factor. It lacks an elaboration of risk, liability and even places of delivery. The unipolarity implies that the price is set by one side and the other side has no choice but to accept. The single-sale certainty refers to the inefficient trades and inconvenient resale. The stiffness refers to the small leeway of bargain, humdrum in form and lack of choice. The preference refers to the egoism, the lack of characteristics of proper communication and rebus sic stantibus. It is widely used in the domestic trade due to its simplicity, importance and easiness.

In turn, the pricing system based on trade terms is recessive, structural, bi-directional, chain transactions, flexible, inclusive and popular in foreign trade. The constitutive property refers to the numerous choices of price terms in a single trade. The bi-directionality refers to the bargains in trade so as to reach a compromise. The chain transaction refers to the convenient resale of cargo. The resale is very easy as long as the parties have the same understanding no matter the cargo is imported or exported or the buyer is from home or abroad. The flexibility means that the price is not based on a single point but on a series of points in the system and there should be one point in conformity to the parties, which offers lots of options for the success of trade. The inclusiveness means that the price terms contain lots of information of services, such as transportation, insurance and freight, etc. Unfortunately, the trade terms are only applied in foreign trade. The enterprises dealing in domestic trade either don't know the features of this pricing system or reject it because of path dependence, which results in the seclusion of the home trade nowadays.

5.2 *The comparison between the INCOTERMS 2010 Rules and Contract Law*

First of all, the Contract Law is mandatory and the INCOTERMS 2010 rules are quasi mandatory. There's no conflict between them in the application.

Secondly, the INCOTERMS 2010 rules provide several trade terms representing symbolic delivery, such as FOB, CFR, CIF, FCA, CPT, and CIP. They can work in the string sales done by the middlemen to make the cargo and capital flow conveniently. The trade terms will be protected by Contract Law if they are accepted by the parties. The delivery way in the Contract Law is inflexible since there's only an actual delivery demanding the cargo and documents to be delivered simultaneously. While the several actual delivery rules in the INCOTERMS 2010, such as EXW, FAS, DAT, DAP and DDU, are elaborated more than that in Contract Law.

Thirdly, the INCOTERMS 2010 rules provide a very precise and detailed description about the place of delivery. For example, the place of delivery in EXW refers to the places storing goods, such as warehouse, factory, etc. where delivery takes place and completes. The place of delivery in FAS refers to the named port within reach of a ship's lifting tackle. The place of delivery in FCA refers to the place where the cargo is transferred to first carrier nominated by the buyer. If the shipping point is the place where the contract is signed, the seller has the obligation of loading (loading the goods on the means of transport provided by the buyer); if the shipping point is not the place where the contract is signed, the seller has no obligation of loading (the seller has no obligation to unload the goods from the seller's means of transport and the delivery is completed when the goods are placed at the disposal of the carrier or another person nominated by the buyer). The place of delivery in DAT refers to the terminal at the named port or place of destination. The seller must unload the goods from the arriving means of transport. The place of delivery in DAP refers to the named place agreed by the parties. The seller delivers when the goods are placed at the disposal of the buyer on the arriving means of transport ready for unloading at the named place of destination. The seller bears all risks involved in bringing the goods to the named place. The provisions on place of delivery in Contract Law are more ambiguous than those in the INCOTERMS 2010 rules, which often result in disputes among the parties[19].

Fourthly, the INCOTERMS 2010 provide precise and detailed rules on the seller's charges. It not only has general rules, but also has rules on specific affairs, such as carriage contract, insurance contract, inspect and pack of goods, tariff, taxation, and the charges of mandatory inspection before shipment, etc. Therefore, if the parties can incorporate the INCOTERMS 2010 rules into their contract of sale, they will have a clear idea on their obligation including burden of charges and reduce the disputes at a maximum degree.

Finally, the INCOTERMS 2010 provides flexible and practical rules on terms of payment. It's a justified convention for the buyer to pay the price of goods. It's unnecessary to emphasize the payment of goods in code. Yet Contract Law provides very detailed provisions on the payment of goods, such as amount, date and place. No matter how precise the contract is, it can not get rid of the dishonest behavior.

In the practice, especially in buyer's market, it's not the contract but the regulations and process reengineering to make the seller deliver goods and the buyer pay the price without the worry of expediting or back money. The terms of delivery and payment in international trade outclass that in domestic trade. The terms of payment in foreign trade are mainly L/C, remittance and collection, all of which have precise date, place and way though they are based on different basis of credit. In domestic trade, the parties usually adopt cash and bank transfer as terms of payment. Though there're provisions on terms of payment in the contract, there's credit sale or back money among the channel distributors, especially those in powerful position, which makes the contract just like a mere scrap of paper. The existence of the dead loan, repudiation affects the normal flow of the domestic goods and the sound development of market economy. By contrast, the intervention of intermediary or a third party can prevent expediting or back money taking place in international trade, resist the ubiquity of forged and fake commodities and make the trade more secure[20]. Furthermore, the application of trade terms can stimulate the reform the mode of payment in domestic trade.

6 CONCLUSION

The integration of foreign and domestic trade is a basic national policy based on market mechanism.

Today, it's essential to have a deeper understanding of WTO rules in circulation area so as to "guide the actions of market players and promote domestic trade to develop sound and fast". The enterprises dealing in domestic trade and foreign trade will focus on different aspects in the systematic project of integration of domestic trade and foreign trade. If the enterprises dealing in foreign trade focus on the internalization of international trade, domestic trade focus on the internationalization, this should be the highlight in the integration of foreign and domestic trade. The application of the trade terms in the INCOTERMS 2010 rules will help to solve the "glass ceiling" caused by the split of foreign and domestic trade, to allocate the resource rationally, and to stimulate enterprises' vitality. These will benefit the upgrading and acceleration of the domestic trade, and it will help the domestic trade meet the international market and the enterprises dealing in foreign trade enter the domestic market.

As price terms, trade terms can be incorporated into product pricing of marketing. It has three functions. First of all, it can stimulate the enterprises to take an active part in the integration project; secondly, it can offset the flaws of marketing pricing theory; thirdly, it can simplify the procedure of business negotiation to save cost and increase efficiency. Accordingly, it's vital for the enterprises dealing in domestic trade to incorporate the INCOTERMS 2010 rules into their trade, change business practice, adjust industrial structure and promote enterprise competitiveness. It's predictable that the application of trade terms in the INCOTERMS 2010 rules is inevitable for the enterprises in the future, so does the reform in circulation area, such as transport, insurance, methods of payment and commodity inspection etc.

NOTES

This article is a substantive result of 2014' Hunan province decision-making and counseling research project: "Research on Accelerating the Development of Modern Service Industry in Hunan – Based on the Analysis and Improvement of Business Flow Level in Commodity Circulation" (No. 2014BZZ078).

REFERENCES

[1] Zu-yi Tan. 2011. Conceptual framework for integration of internal and international trade and its market operation mechanism, *Business Research* 2011(4): 90–95.

[2] Jun-fa Ding. 2004. Integration of foreign and domestic trade and innovation of circulation, *Marketing Guide* 2004(3): 6–8.

[3] Xi-lan Zou & Jia-long Tan. 2013. Depth testing for the integration of foreign and domestic trade in Guangdong province, *China economic weekly* 2013(25): 58–59.

[4] Yu-lin Zhang. 2008. The integration of foreign and domestic trade: A book without words from Heaven, *International Business Daily*, T26, 12-18-2008.

[5] Zi Ye. 2011. Promoting the integration of domestic and foreign trade: Abutting joint of distribution channels is the key, *Consumption Daily*, A03, 12-29-2011.

[6] Tao Hong & Zhen-rong Zhu. 2013. Joint development of domestic and foreign trade (eds), *Economic Management Publishing Press*, Beijing, China: 1–11.

[7] Yong-jing Zhang. 1998. Significance of 'internalization of international trade' in China's open economy, *International Economics and Trade Research* 1998(6): 25–27.

[8] Jing-fei Ran. 2004. Domestic and international trade studies on co-integration and causality, *International Trade Journal* 2004(12): 10–14.

[9] Ling-li Chen. 2003. Talking about the application of trade terms in domestic trade, *Northern Economy and Trade* 2003(12): 10–11.

[10] Ting Zhou. 2006. See the domestic trade application of international trade terms from a case, *Practice in Foreign Economic Relations and Trade* 2006(5): 33–34.

[11] Ting Zhou. 2006. Discuss a character in 'domestic trade' of the INCOTERM EXW, *Practice in Foreign Economic Relations and Trade* 2006(10): 36–38.

[12] Xin Wu. 2009. Analysis in main mode and nature of the transaction of EXW Incoterms in our country, *Modern Economic Information* 2009(14): 13.

[13] Jian-zheng Ge. 1998. Drawing on international trade terms to explore coal trading rules, *North China Power* 1998(9): 29–31.

[14] Yu Peng. 1977. Evolution of FOB Contract, *International Trade Journal* 1977(3): 15.

[15] Jiang-nan Zhai. 2012. Theory and practice of trade in goods (2nd ed), *University of international business and Economics Publishing House*. Beijing, China: 45.

[16] De-ming Chen. 2009. Several issues relating to domestic and foreign trade, *Seek the Truth* 2009(7): 21–24.

[17] Jiang-nan Zhai. 2012. Research on the application of Incoterms 2010 in domestic trade in China, *ICEEM 2012 colloquia Edited by Hong Kong Education Society* 2012(5): 697–700.

[18] Jiang-nan Zhai. 2012. New technology and marketing innovation-base on leading trade terms into pricing system, *MSIT2012 Edited by ST. PLUM-BLOSSOM PRESS PTY LTD*, Melbourne-Australia 2012(6): 86–91.

[19] Da-peng Li. 2001. Complementary two sets of rules, *Journal Chongqing Institute Commerce* 2001(4): 31–33.

[20] Hua Zhu. 2012. Integrating domestic and international trade of shanghai, *World market* 2012 (Z2): 22–27.

Industrial Engineering and Management Science – Lee (Ed)
© 2015 Taylor & Francis Group, London, ISBN 978-1-138-02647-6

The efficiency of China's urbanization

JunQi Li & HongWei Zhang
Sichuan University, Chengdu, China

ABSTRACT: China's urbanization has made great advances, but there are a lot of problems. In general, the urbanization still stays in the extensive stage. In order to achieve intensive development of urbanization, it is necessary to study the efficiency of urbanization to find out reasons for restricting the efficiency of urbanization and countermeasures. With the help of DEA model, this paper evaluates the efficiency of China's urbanization from 2003 to 2012. The convergence analysis and the analysis based on Malmquist index of the evaluation results are made. According to the empirical results, this paper proposes countermeasures.

Keywords: urbanization, efficiency, resources, systems, reforming

1 INTRODUCTION

By introducing reform and opening up, China has greatly emancipated productivity. One remarkable achievement is to introduce the urbanization. According to the data of China Statistical Yearbook, the proportion of urban population in the total population rose from 17.92% in 1978 to 52.57% in 2012. The largest population migration in the history of mankind occurred in China. The tremendous development of urbanization is promoting the social and economic progress. However, this achievement comes with the problems of environmental pollution, low residential income, low consumption propensity, urban-rural dualistic structure and so on. In general, the urbanization in China stays in the extensive stage of development. Correspondingly, it is urgent to change from the extensive urbanization to the intensive stage. Subsequently, the measurement of urbanization efficiency plays a critical role in developing the effective urbanization in China. And the measurement and corresponding strategies carry a strong practical significance.

There are two inadequacies in the current research of urbanization developments: one is enormous system of indicators, which makes heavy demands on the availability of data. The enormous indicator system also complicates the evaluation of results, which can hardly make horizontal or longitudinal comparisons. The other is that the data unavailability hinders the DEA approach extending to the research covering the whole of China. This paper builds an indicator system to measure the efficiency of urbanization in China based on the available data. By adopting the DEA-BC^2 model, the paper studies the quantitative evaluation of China's urbanization efficiency from 2003 to 2012. In the model, the DEA efficiency represents the urbanization efficiency.

2 DEA MODEL

DEA model, data envelopment analysis, was first set up by A. Charnes and W.W. Cooper in 1978. The DEA model is an interdisciplinary subject, involving many branches of mathematics, operations research, management and mathematical economics (Wei Quanling, 2004).

It is an important estimation method of nonparametric statistics based on the models set by Farrell. The DEA model constructs the production possibility set to assess the relative effectiveness by analyzing the multiply input-output decision making units.

DEA-BC^2 model from the series of DEA models is based on the assumption of Variable Return to Scale for each decision making unit. This assumption was first proposed by Banker, Charnes and Cooper in 1984. The technical efficiency (TE) in this model could be decomposed into the product of pure technical efficiency (PTE) and scale efficiency (SE). Specifically, the pure technical efficiency demonstrates the production efficiency determined by the management and technique in terms of constant scale. The scale efficiency shows the production efficiency driven by scale factors. Additionally, the technical efficiency is also known as comprehensive technical efficiency, which is the comprehensive evaluation of efficiency regarding resource allocation and utilization by decision making units.

3 THE EMPIRICAL ANALYSIS OF URBANIZATION EFFICIENCY IN CHINA

3.1 *Variables and data*

3.1.1 *Variables*

First of all, it is critical to select the proper variables of input and output for assessing the urbanization

Table 1. The descriptive statistical characteristics of each variable from 2003 to 2012.

	urban population	non-agricultural output	urban fixed asset investment	urban employment	permanent resident population
Mean	1975.72	7365.80	4143.07	615.54	4224.68
Standard Deviation	1325.62	7051.34	3792.66	363.06	2676.41
Minimum	64.89	143.80	133.96	47.67	270.17
Median	1678.22	5302.64	3045.29	576.09	3767.00
Maximum	7140.36	36342.74	22248.58	1557.42	10594.00

Sources: China Statistical Yearbook from 2004 to 2013.

efficiency in China from 2003 to 2012. The followings are the guidance of the variable selection. Firstly, the variables should meet the requirements of economic implications. Secondly, the selection of variables should allow the availability of data. However, this guidance might generate the potential limitations of the model setup and final evaluation results. Finally, the variable selection should satisfy the requirements of DEA model itself, which include the positive value, isotonicity between input and output value. On the basis of the above guidance, it turns out to be accurate to use urban population and non-agricultural output value as the output variables, accordingly, urban fixed asset investment, urban employment, permanent resident population as the input variables[1]. The reasons are as follows: The main urbanization rate used in China is the proportion of urban population in permanent resident population. Consequently, urban population is selected in this paper as output variable while permanent resident population is input variable. In addition, the paper selects urban fixed asset investment, urban employment as input variable and non-agricultural output value as output variable. This selection demonstrates the dependency of output on capital and labor. Theoretically, urban fixed asset investment and urban employment has positive influence on the increase of urban population. In the meanwhile, the permanent resident population may also positively affect the non-agricultural output value. The paper will conduct the correlation test to prove the above assumptions.

3.1.2 *Data*

The followings are the descriptive statistics of 31 provincial-level administrative regions in China from 2003 to 2012. The data are from China Statistical Yearbook from 2004 to 2013. The variables are urban population (ten thousand), non-agricultural output value (One hundred million yuan), urban fixed asset investment (One hundred million yuan), urban employment (ten thousand), permanent resident population (ten thousand).

[1]non-agricultural output = GDP-agricultural output. Urban fixed asset investment, GDP, agricultural and non-agricultural output are all based on the calculation of data which take 2003 as base period.

Table 2. The Pearson correlation test of input-output.

	urban fixed asset investment	urban employment	permanent resident population
urban population	0.740**	0.926**	0.903**
non-agricultural output	0.883**	0.810**	0.701**

Notes: **is significantly correlated in the level of 0.01.

3.2 *Correlation test of variables*

The input and output variables must meet the requirements of statistical isotonicity. Accordingly, the correlation test is conducted by SPSS19.0 with the following results.

Based on the results of correlation test, input and output variables meet the requirements of statistical isotonicity. Therefore, DEA model is proper to be adopted to assess the efficiency of decision making unit.

3.3 *The assessment of urbanization efficiency on DEA-BC2 model: the evidence from 2003 to 2012*

Here is the result of the urbanization efficiency run by the DEAP2.1 based on the data set of 31 Chinese provinces from 2003 to 2012.

According to the above table, the urbanization efficiency of each province in China has a certain positive correlation with its own state of economy from 2003 to 2012. Yunnan and Guizhou are remarkable exceptions in the western provinces. The mean efficiency over 10 years respectively reaches 0.966 and 0.915, which is generated from their less developed agriculture. Therefore, the non-agriculture holds a large proportion of the gross domestic product, causing the population mobility.

3.4 *Convergence test of urbanization efficiency*

Based on above analysis, 31 provinces vary greatly in the urbanization efficiency. It is necessary to gain more insight into the fluctuant trend of efficiency variation.

Table 3. The mean and decomposition of the urbanization efficiency from 2003 to 2012.

Province	TE	PTE	SE	Province	TE	PTE	SE
Beijing	0.967	0.981	0.986	Hubei	0.771	0.809	0.959
Tianjing	0.942	0.990	0.951	Hunan	0.825	0.855	0.967
Hebei	0.818	0.831	0.986	Guangdong	1.000	1.000	1.000
Shanxi	0.770	0.800	0.962	Guangxi	0.878	0.900	0.974
Neimenggu	0.744	0.787	0.945	Hainan	0.858	0.991	0.865
Liaoning	0.844	0.855	0.987	Chongqing	0.883	0.916	0.963
Jilin	0.794	0.819	0.968	Sichuan	0.862	0.884	0.977
Heilongjiang	0.898	0.907	0.989	Guizhou	0.915	0.960	0.951
Shanghai	1.000	1.000	1.000	Yunnan	0.966	0.991	0.975
Jiangsu	0.947	1.000	0.947	Xizang	0.394	1.000	0.394
Zhejiang	0.979	0.989	0.990	Shaanxi	0.732	0.755	0.968
Anhui	0.823	0.837	0.983	Gansu	0.765	0.827	0.920
Fujian	1.000	1.000	1.000	Qinghai	0.649	0.989	0.656
Jiangxi	0.702	0.716	0.979	Ningxia	0.702	1.000	0.703
Shandong	0.849	0.934	0.909	Xinjiang	0.646	0.735	0.880
Henan	0.786	0.796	0.988	Average	0.829	0.898	0.926

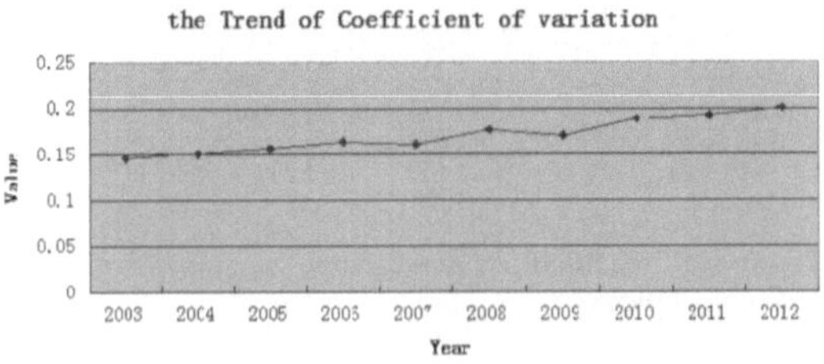

Figure 1. The trend of coefficient of variation-δ convergence.

It carries a significant meaning to narrow the regional economic development disparity. Barro (1991) maintained underdeveloped countries with less capital per capita would gain a higher marginal capital output than the developed countries based on the assumption of the diminishing marginal returns of capital in terms of the neoclassical growth model. The output in less developed countries would accelerate in a higher speed and narrow the gaps among the economic growth in different countries. Accordingly, Barro and Sala-i-Martin (1991) proposed β convergence and δ convergence. β convergence estimates the convergence trend by analyzing the economic growth rate while δ convergence diagnoses the convergence trend by analyzing the economic growth level. The former is the necessary but not sufficient condition for the latter.

δ convergence analysis estimates the convergence of samples through calculating their standard deviation. Due to the significant difference among the mean urbanization efficiencies in 31 provinces, it is unable to use the standard deviation to conduct δ convergence. Therefore, coefficient of variation is the substitute for standard deviation. Once the coefficient of variation demonstrates a gradual declining trend with time in a period, the δ convergence exists. Otherwise, the samples is diverged. The following figure shows the trend of the coefficient of variation.

From Figure 1, it is obvious that there is no δ convergence in China's urbanization efficiency over years. In other words, the differences of the urbanization efficiency grow larger among the provinces in China, indicating the deteriorating uneven development of China Urbanization. This should be put more attention to.

The empirical result shows the analysis of β convergence hardly work out because the fitting outcomes hardly pass the statistical test. Therefore, The test of β convergence will not be conducted.

3.5 *The analysis of the urbanization efficiency change based on the DEA-Malmquist model*

Over the same period, each decision making unit is still in the same environment with equivalent technique and systems. As time goes by, the technique and systems might change, leading the production frontier to move. The DEA is a static method based on cross-sectional data. The passage of time would lead to the production frontier move, causing the inability for traditional BC^2 model to process the time-series data. Therefore, it is hard to compare the efficiency of decision making unit at different time. The adoption of Malmquist index approach could easily tackle this problem. One benefit of Malmquist index is that this approach could reveal the movement of the production frontier caused by the time. So it results in the longitudinal comparison among the decision making units. Malmquist index approach was first set up by Sten Malmquist in 1953. Malmquist index approach was adopted by Caves etc. to measure the productivity in 1982 and was widely used thereafter. The combination of Malmquist index and DEA generated the DEA-Malmquist index, which was proposed by Fare etc. in 1992. The definition of Malmquist Index is shown as following by the introduction of distance function.

$$M(x_t, y_t, x_{t+1}, y_{t+1}) = \sqrt{\frac{D_{t+1}(x_{t+1}, y_{t+1})}{D_{t+1}(x_t, y_t)} \times \frac{D_t(x_{t+1}, y_{t+1})}{D_t(x_t, y_t)}}$$

(x_t, y_t) and (x_{t+1}, y_{t+1}) respectively demonstrate the combination of inputs and outputs at time t and t + 1; D_t and D_{t+1} respectively represent distance function at time t and t + 1 with the reference to the technique at the time t. $\frac{D_t(x_{t+1}, y_{t+1})}{D_t(x_t, y_t)}$ and $\frac{D_{t+1}(x_{t+1}, y_{t+1})}{D_{t+1}(x_t, y_t)}$ respectively represent Malmquist Index with the reference to the technique at time t and t + 1. To avoid the differences caused by the random choices, the two indexes are conducted by geometric mean, obtaining the above model. It can be further decomposed into:

$M(x_t, y_t, x_{t+1}, y_{t+1}) = \text{TFPCH} = \text{TECHCH} \times \text{EFFCH}, \text{EFFCH} = \text{PECH} \times \text{SECH}$

TFPCH is the total factor productivity change index, which represents the change of total factor productivity. In this paper, TFPCH refers to the change of urbanization efficiency, revealing the fluctuation of urbanization efficiency in the adjacent two years; Secondly, TECHCH[2] is the technical progress index, reflecting the movement of the production frontier caused by the technical and system factors. Thirdly, EFFCH is the technical efficiency change index, which reveals the change of urbanization efficiency caused by the management, technique and scale factors of the decision making units in the face of constant production frontier. Specifically, the change is shown by the resource allocation and utilization. Fourthly, PECH is the pure technical efficiency change index which shows the impact by management and technical factors on urbanization efficiency. Subsequently, SECH is the scale efficiency change index and reflects the influence on the change of the urbanization efficiency by scale factors. Obviously, Malmquist Index Approach shows not only the change of technical efficiency, but also movement of the production frontier. The Malmquist Index offers an approach of estimating dynamic DEA efficiency. Those indexes mentioned above take 1 as the cut-off point. The value bigger than 1 means the efficiency is improved compared to the previous year. The value of 1 indicates unchanged efficiency while the value smaller than 1 refers to the deteriorating efficiency. The quantitative results operated by DEAP2.1 is given as following according to the Malmquist Index for each province in China from 2003 to 2012.

Here is trend as following based on the quantitative results of Table 4.

In the light of Figure 2, the trend of the total factor productivity change index (TFPCH) demonstrates there is a declining trend in urbanization efficiency for China from 2003 to 2012. Although the rate of decline slowed during 2009 to 2012, the overall situation is not optimistic. The decomposition of TFPCH shows TECHCH and EFFCH are larger than 1 during 2009 to 2010 and from 2010 to 2011 respectively, and smaller than 1 in the rest of the years. There is a general declining trend of technical efficiency and production frontier. The decrease of the technical efficiency reflects the resource waste and distorted allocation in the process of urbanization. The declining production frontier shows the system environment is not satisfactory for the development of urbanization in China. The problems of household registration system, land system, employment system and social security system have hindered the development of urbanization. More specifically, Chinese government plays a critical role in the infrastructure investment, and also determines the allocation of land and financial resources. Government and state-owned enterprises enjoy priorities in the allocation of land and financial resources, which crowd out the private investment. The low efficiency of the state-owned sector restricts the improvement of the urbanization efficiency. There are urban-rural dualistic characteristics of current household registration system, employment system, land system and social security system. They are the main obstacles to the urbanization.

Each index experienced significant fluctuation during 2008 to 2012. The fluctuation was caused by the China's 4 trillion yuan ($587 billion) stimulus package, as well as the reform of the systems. In detail, the sharp increase of TECHCH indicated the optimizing system environment, which mainly involved the reform of land system. *Decision of the CCP Central Committee on rural reform and development* passed on October 12, 2008 proposed to reform the land system, which partly recognizes the circulation of land use rights of peasants. Peasants can obtain partial

Table 4. The mean and decomposition of the Malmquist index.

Year	EFFCH	TECHCH	PECH	SECH	TFPCH
2003–2004	0.991	0.955	0.997	0.994	0.947
2004–2005	0.974	0.963	0.975	0.999	0.938
2005–2006	0.982	0.966	0.977	1.005	0.948
2006–2007	0.996	0.968	1	0.996	0.965
2007–2008	0.983	0.989	0.991	0.992	0.973
2008–2009	0.982	0.948	0.99	0.993	0.931
2009–2010	0.939	1.051	0.952	0.987	0.987
2010–2011	1.013	0.957	1.011	1.002	0.969
2011–2012	0.996	0.994	0.989	1.008	0.991
Average	0.984	0.976	0.987	0.997	0.961

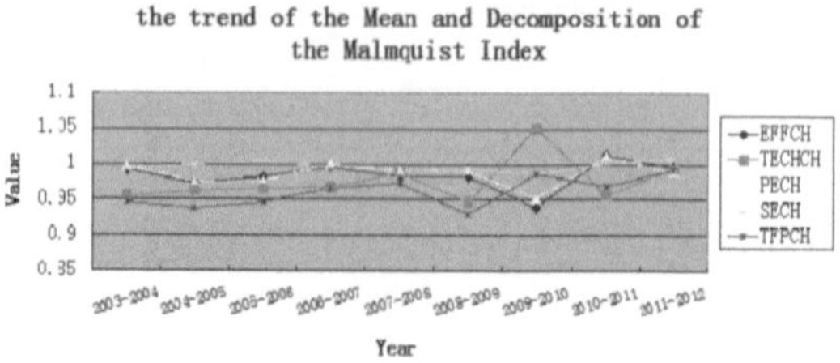

Figure 2. The trend of the mean and decomposition of the Malmquist index.

[2]Due to the limited time interval, the change of technique is not significant. Consequently, TECHCH mainly reflects the movement of production frontier determined by the systems.

benefit of the circulation. This reform breaks the system barriers to the intensive management of land, and lays the foundation of further optimizing urbanization efficiency. The *decision* satisfies the demand of land during the process of urbanization and allows more peasants integrate into the urbanization. EFFCH fluctuated greatly during 2009 to 2011 mainly because of the China's 4 trillion yuan ($587 billion) stimulus package. EFFCH declined sharply from 2009 to 2010. It illustrates the sudden increase in the fixed asset investment resulted in further deterioration of the situation of the resource waste and distorted allocation. Fortunately, this situation improved slightly during 2010 to 2011.

Moreover, it is obvious that 2007 is a turning point. Before 2007, TFPCH has the similar trend with EFFCH. After 2007, TFPCH changed in the same pattern with TECHCH. This shows that the main factor to change urbanization efficiency was the change of technical efficiency before 2007 while the main factor after 2007 was the movement of production frontier. This change reveals urbanization efficiency remained low primarily because of the resource waste and distorted allocation before 2007. After 2007, the main obstacles became the problems of the system environment. It is distinctly shown by the deviation of TFPCH and EFFCH from 2009 to 2011. At the beginning of Reform and Opening up, the urbanization in China stayed in the extensive stage when the rapid development was primarily driven by the resource investment. However, when the urbanization developed into a certain level, the successive input resulted in the waste of resources. The main driver became the system reform. Consequently, 2007 is the turning point of China urbanization from the extensive stage to the intensive stage.

The trend of EFFCH is also a sobering picture. According to Figure 2, SECH run smoothly around 1, showing EFFCH is mainly driven by PECH. Therefore, the management of the government had a negative impact on the technical efficiency. This resulted in further deterioration of the situation of the resource waste and distorted allocation. In China, government mainly determines the social and economic development as the administrator. They are responsible for the allocation of resources. The great fluctuation of PECH during 2009 to 2011 brought the fluctuation of EFFCH. This fluctuation was outcome of low efficiency of resource allocation set by the government, who conduct the China's 4 trillion yuan ($587 billion) stimulus package.

4 CONCLUSION

Through empirical analysis, We draw the following conclusions. Firstly, the urbanization efficiency of each province in China has a certain positive correlation with its own state of economy. The difference of urbanization efficiency is remarkable, and always expanding. Secondly, nationwide, the overall urbanization efficiency experiences a trend of annual declines although the rate of declining efficiency slowed down after 2009. Thirdly, there are two major factors that led to the low efficiency of urbanization: the resource waste and distorted allocation, as well as the problems of the system environment. The former was the main driver of low urbanization efficiency before while the latter was main driver after 2007. Fourthly, each index experienced significant fluctuation during 2008 to 2012, which was caused by the China's 4 trillion yuan ($587 billion) stimulus package, as well as the system reform. The reform of land system promoted the urbanization efficiency while the China's 4 trillion yuan ($587 billion) stimulus package resulted in further deterioration of the situation of the resource waste and distorted allocation.

Some advices are given based on the above analysis as followings: China should promote the urbanization efficiency by changing the government roles and reforming the systems. On the one hand, the market mechanism should be emphasized with less policy interventions, especially in the resource allocation. On the other hand, the further system reform should be carried out to eliminate the discrimination in household registration system, land system, employment system and social security system with the ultimate goal of eliminating the urban-rural system disparity.

REFERENCES

Barro R J. Economic growth in a cross section of countriesc [J]. The Quarterly Journal of Economics, 1991, 106(2): 407–443.

Barro R J, Sala-i-Martin X. Convergence across states and regions [J]. Brookings papers on economic activity, 1991: 107–182.

Caves D W, Christensen L R, Diewert W E. The economic theory of index numbers and the measurement of input, output, and productivity [J]. Econometrica: Journal of the Econometric Society, 1982: 1393–1414.

Wang Xin. Total factor productivity measurement of China's Equipment Manufactoring industry [D]. Chengdu: Southwestern University of finance and economics, 2011.

Wei Quanling. Data Envelopment Analysis [M]. Beijing: Science Press, 2004.

Industrial Engineering and Management Science – Lee (Ed)
© 2015 Taylor & Francis Group, London, ISBN 978-1-138-02647-6

The problems and countermeasures of the logistics financial business of city commercial banks in China

HuaLan Lu
Economics College, Sichuan University, Chengdu, Sichuan, China

ABSTRACT: Logistics financial business has been rapid development in foreign countries, especially in the U.S. Relative to the foreign business, this business in China started late, being just at the stage of beginning development. Because of the development pattern of the logistics financial business and the existing problems of city commercial banks, more and more banks will be put into our market positioning and competitive strategy system. This is bound to become the new profit growth point. Standing on the perspective of bank management, this article analyzes the optimized development scheme of the logistics financial business of the city business in the future.

Keywords: Logistics financial business; profit growth point; risk; mode; city commercial banks

1 INTRODUCTION

1.1 *The development background of the logistics financial business*

Logistics financial business is the combination of finance and logistics, which was developed in 2400 BC, when the Mesopotamian area saw grain warehouse receipts. And in the earliest UK, the circulation of paper money was silver warehouse receipt of payment. Logistics business in China experienced rapid development since China's accession to the WTO. It was once considered by companies as "the third profit source", and grew at an annual rate of 16% to 25%. This is mainly the warehousing management pattern of traditional enterprises, such as the national large-scale storage and transport enterprises and sino-foreign joint venture enterprise. However, as the market competition is increasing fiercely and the third party logistics enterprise's profit has dropped to 2%–3%, the enterprise faced an uncertain future. In this situation, many logistics companies began to explore the third party logistics business model and put it into use. Logistics service gradually extended to other aspects of the value chain such as finance, e-commerce, marketing and other derivative services. Among them, the logistics financial service is the combination of logistics flow and cash flow, which is one of the important logistics derivative services.

With the financial market competition intensifying, in order to seek new profit growth point, banks are scrambling to study and introduce new products around the micro, small and medium enterprises. This practice accelerated the structure transformation. Then logistics financial business also arose at this historic moment. However, on the whole, it was relatively late for Chinese banks to develop logistics financial business, and also it was not perfect, immature and standard. There are more loopholes and operation risks, especially in the city business of some small and medium-sized banks. And domestic study on logistics financial business is far from enough, especially from the perspective of the bank. The research on city business development of the logistics financial business is still blank. In this case, this paper stands in the angle of the bank to study about logistics financial business, especially in the problems and countermeasures.

1.2 *The present situation of the logistics financial business in China*

1.2.1 *Logistics enterprises themselves carry out financial business*

In foreign countries, the logistics financial business is the financial institutions. While in China, it is logistics enterprises. According to statistics, the number of the listed existing logistics enterprises of our country is more than 40, including ocean logistics and freight logistics etc. At present, the China material storage and transportation corporation (CMST) carried out the earliest and largest logistics financial business. It is the member of the department of China Chengtong group (CCT) and the nation's largest warehousing enterprises. It carried out the warehouse receipt pledge business in 1999. CMST owns mature management experience and has the advantages of its own domestic warehouse network throughout the major cities. Besides, its parent company is in China. Therefore, it has explored a variety of warehouse receipt pledge financing business regulation mode at present.

1.2.2 *Logistics enterprises and the bank begin to cooperate*

Logistics enterprises control the whole industry chain and create a new growth space through the deep cooperation with banks involving the logistics financial business. It is also beneficial to form its own competitive advantage, maintain customer relationship and provide customers finance etc. All of these greatly improve the position of logistics enterprises in the customer heart and attraction. For example, in July 2006, the industrial and commercial bank of China and foreign trade transportation corporation of China signed 'the logistics financial strategy framework agreement'. In this agreement, the two agreed in supervising financing and logistics, logistics and settlement, logistics factoring, logistics guarantee and customers' credit risk management and so on. Its purpose is to develop new products, connect both sides service platform and extend the range of customer service. These improve their efficiency, reduce the cost and increase the value-added services, etc. So it is imperative for the logistics enterprises and bank cooperation to cooperate. Also it is a win-win cooperation model. Now CMST has signed a lot of total framework agreements with more than 20 banks to develop logistics financial business, including industrial and commercial bank, the bank of China, bank of communications, China everbright bank, China Minsheng Banking Corp. Ltd, Citic bank, Huaxia bank, China merchants bank, French Paris bank, standard chartered bank, Austria's central bank, etc.

1.2.3 *Logistics financial business in China's banking industry was carried out later*

By 2013, China has 17 major commercial banks carrying out the logistics financial business, and the scale is gradually expanding. Other small and medium-sized Banks and local commercial Banks are exploring and introducing gradually logistics financial business products. Nevertheless, the understanding of the logistics financial business and its products for China's banks only stay in the traditional mode. The depth in understanding of the logistics financial business level differs from bank to bank, so the effect is also different. But most Banks have got more profitable. In 1999, SDB began to explore the goods and business. In 2001, it launched personal property and goods right pledge business. In 2004, it began the agent discount business. In 2005, it made trade finance and small and medium-sized enterprise's strategic transformation come true. While between 2006 to September 2008, "SDB supply chain finance" brand was introduced. And SDB created regulation for the logistics enterprises and logistics value accumulated nearly 800 billion yuan. Huaxia bank was unwilling to be left behind. On September 7, 2007, it introduced its new building logistics financial business product in Chengdu, Sichuan. At the same time, it signed with Suning, Yurun, five-star and CMST etc. which are well-known enterprises about the logistics financial business cooperation. In 2009, China citic bank in the business of supply chain finance raised more than $300 billion. Today, China Everbright bank, China Minsheng Bank, the Construction Bank, China Merchants Bank also regard the logistics financial business as a focus on the development of business. However, in the aspect of logistics financial business and products, China's banking industry is still in its infancy stage, and needs constant innovation and development to win the market better.

The logistics financial business started later for city commercial banks, far behind the big Banks and joint-stock Banks. Among them, the Chongqing Bank began to explore warehouse receipt pledge and silver ticket business until 2005.

2 THE BARRIERS DURING CARRYING OUT THE LOGISTICS FINANCIAL BUSINESS FOR CITY COMMERCIAL BANKS

2.1 *Backward system*

City business banks were established on the basis of urban credit cooperatives in the mid of 1990s. City credit cooperatives were established for the tertiary industry and some financial institutions. It aimed to adapt to the needs of employment for children at the beginning of the reform and opening-up in China. System, mechanism, management level, personnel quality, etc. have congenital deficiencies, urban credit cooperatives accumulated a lot of risk during the development. In order to dissolve the risk, the state council decided that more than 30 large and medium-sized cities across the country gradually formed city business banks in 1995. Until the November of 2013, city commercial Banks increased to at least 145. So, from the point of development, city business bank are not a real bank. The city firms system is relatively backward. The result is the starting and development of the logistics financial business is relatively lagging behind. The system needs a series of reforms, and this is a long process full of pain. Pepsi is to be new. So, it is probably too busy to take care of some business research and innovation. Then the understanding of the logistics financial business and new business varieties certainly are not enough.

Chongqing bank, for example, is compared with the joint-stock Banks: their balance of companies and logistics financial business respectively in 2010 and in 2012 are shown in figure 1 and figure 2.

2.2 *There is no professional talents and specialized management institutions*

Our banks talents proficient in logistics, finance, legal, and innovation ability are far *cry* from the developed countries. There are not enough talents proficient in logistics financial business knowledge and advanced financial knowledge, which seriously influences the innovation ability and profitability of the bank. So does the nCity business banks. Most city businesses banks, such as the Baoshang bank, bank of Deyang,

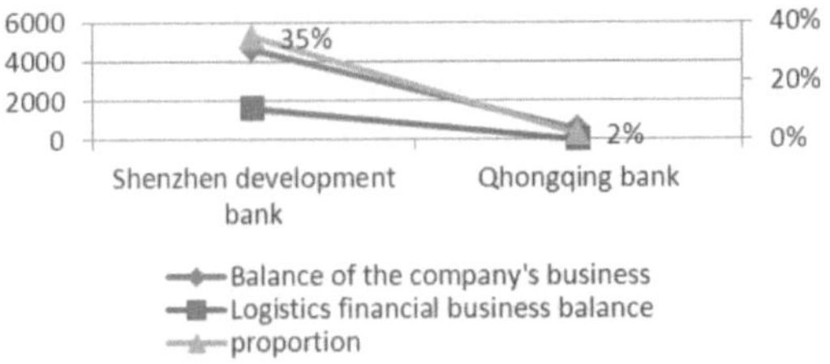

Figure 1. The comparison of logistics financial business balance in 2010.
Units: one hundred million yuan.
Data sources: The site of Ping an bank and the bank of Huaxia.

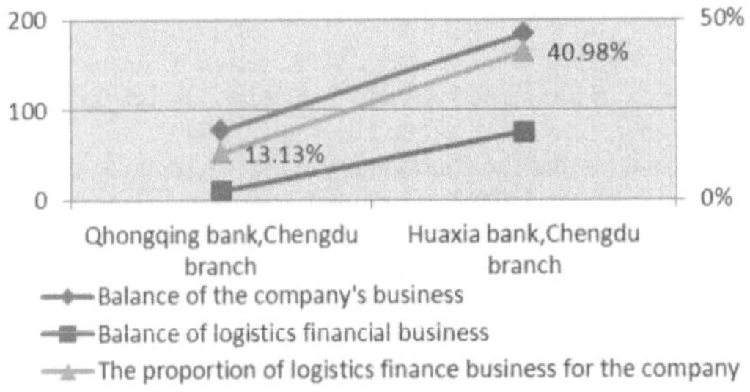

Figure 2. Balance of logistics financial business in August 2012.
Units: one hundred million yuan.
Data sources: The site of Ping an bank and the bank of Chongqing.

Panzhihua bank, Harbin bank, etc., failed to break the personnel mechanism. They introduce talents, especially the professionals, inject fresh "blood" at the beginning of their establishment. Their researchers mainly come from the original urban credit cooperatives who are rigid, backward thinking. Their innovation consciousness is not strong, and the service quality is poor. There is no professional system of logistics finance business research and development. Nor the logistics financial business is on the important agenda. City firms not only started late, but also didn't have system, which just only pruned down on the basis of the others. They had no innovation. So they not only had eliminated some of the important and key elements, but also lost the risk and the significance of development.

At the same time, most of the city business banks have not set up specialized management institutions. For example, the head office of Chongqing bank has not set up the logistics financial business management department. The branches have not set up goods detain centers, which are mainly managed by the company department of headquarters and the company business department of the branches. The concrete management and operation are performed by the credit and business operation entity. They can't do the elaborating management, which virtually makes the promotion and development of the logistics financial business discount. So the main drawbacks followed. Firstly, the management is lack of specialization and refinement. There are about 3 to 5 people in credit and business operation entities. And it is impossible to accord strictly with the requirements of the logistics financial business management. Whenever in the sales department, the headquarters of the company bank or branch company, there are only 1 to 2 people responsible for the management of the logistics financial business personnel, even part-time. This can't satisfy the needs of the development of logistics finance business. Secondly, from the actual situation, most work depends on the credit and business operation entities. In addition to the big workload, management and operation are not separate. So, it is easy to loosen the management. Thirdly, some problems can't be dealt with in a professional, meticulous and timely way, for example, problems in arrying and exchanging audit, tour (nuclear) library, the price audit, marking to market, the price compensation, the calculation of the deposit and the value of the pledged property, parameter of the registration and check. Therefore, possible risks can't be avoided.

2.3 *System is not perfect*

Because it is not long for the logistics financial business to develop in China, the related process specification is not yet perfect. So the development of the logistics financial business has not been fully tapped. The construction of the logistics financial business system doesn't focus on the internal management system of the logistics company, the logistics financial business management system of banks and logistics warehouse management system. From the perspective of the bank, some system defects still exist in the logistics financial business Some banks, especially the city business, don't have the unified standard to accord and the specific professional logistics financial business operation specification. Also it is hard to communicate with enterprises about information to guarantee the implementation of the risk on the handle. And it is not perfect in the regulatory tracking. In addition, the external environment of China's banking industry also can't guarantee banks to implement the interests of the logistics financial business for lack of relevant legal basis in our country. The present law is not completely suitable for the development of the logistics financial business. In the process of handling disputes, the execution efficiency is low while the cost is high, which can't guarantee the benefit of the banks.

2.4 *Lack of innovation*

If every link in enterprise logistics is linked together, including raw materials, spare parts procurement, transportation, processing and manufacturing, distribution, and if the lowest total cost of logistics can be obtained, the bank financing is necessary. Banks can cooperate with every stage of enterprise logistics, for example: accounts receivable, accounts payable, inventory, third-party logistics service providers. As a result, they can design and provide a variety of specific forms of financial business. The logistics financial business of city business banks is at the stage of infancy, conformism, copying, outmoded products, lack of innovation, few varieties, single structure, all of which make it less attractive for customers.

2.5 *Goods and products are not clear*

The choice of the pledge is one of the most important criterion, which decides the logistics financial business' risk coefficient. The value evaluation of pledge is an important content of the logistics financial services, even if it is an enterprise engaged in the work of some commodity warehousing logistics industry long-term and is quite experienced. But certain risk still exits the value evaluation of the pledged goods. The market value of the goods will fluctuate. Some of the goods have strong ability to maintain their value, but some of the commodity price is volatility. If the market price of a commodity causes borrowers not to repay normally because of large degree of volatility, the bank still sold the pledge. The value may be unable to repay bank loans. Thus it brings losses to the bank and bad loan forms.

At present, the city businesses banks do less in pledged product selection. There is no professional system or standard process of logistics financial business to choose the pledge. There is also no clear regulation to decide which goods may be pledged and which can't be pledged. At the same time, there is also no clear standards or specific list of goods that can be pledged. So at the time of shipping enterprise giving the pledge, banks need to reevaluate the value of the pledge to see whether it conforms to the standards and the value lines, etc. This will increase the risk of bank loans at the same time, also reduces the efficiency of bank loans. And, in practice, there is no detailed list to be chosen from, so it brings difficulty to marketing, examination and approval. It also gives the operating personnel the larger decision, which is easy to cause the risk of moral hazard. Or, in order to shirk responsibility, some better pledge are rejected. These arrest the progress of the business. For example, in August, 2012, the pledge breed is not mainly at the bank of Chongqing, Chengdu branch (see figure 2.3), but in the steel industry. In the phase of economic downturn at present, the real estate is in downturn. The iron and steel industry is also in the decline. And a large number of articles focused on the steel industry, which may lead to the concentration risk.

3 OPTIMIZING COUNTERMEASURES OF LOGISTICS FINANCIAL BUSINESS OF URBAN COMMERCIAL BANKS

3.1 *To innovate institution and business*

The logistics financial business is bound to compete for the business profit source of banks. So at this stage the banks should keep pace with times and realize the integration trend of logistics and financial and mutual aid. Besides, they should positively develop all kinds

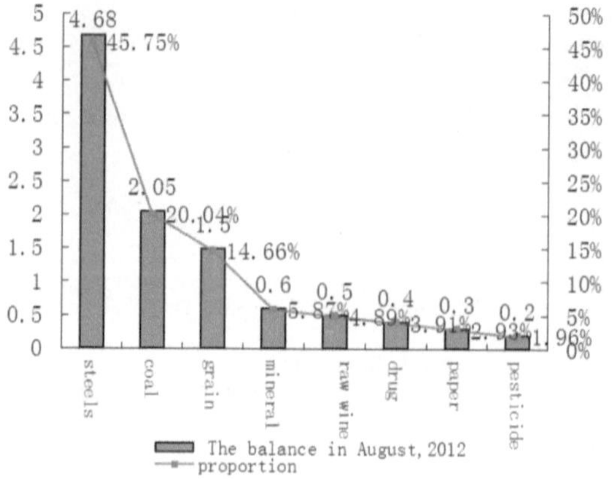

Figure 3. The pledge variety structure of the bank of Chongqing, Chengdu Branch t in August, 2012.
Units: one hundred million yuan.
Source: the web site of Chongqing bank.

of logistics financial business products, grasp the business opportunities and expand the market. The urban commercial banks, in particular, should further deepen the system reform and speed up the change of business development mode. To make themselves "real bank", first, they should improve the understanding of new products. Second, they should make innovative products on the agenda and speed up the exploration, promotion and development of new products. Third, they should fully recognize the significance and functions of logistics financial business. The last but not the least, they should design innovation of logistics financial business operation mode and operation process. Management personnel at various levels should emancipate the mind, change ideas, unify the understanding, make reform heart and soul, grasp the development, consider the logistics financial business to be a new profit growth point to grasp, quickly train it to be a competitive new products in cost line.

To meet the diversity of logistics financial business products, urban commercial banks should carry on the service innovation to encourage business to further expand and develop new business model. Finally, they can meet the needs of customers and the change of environment and the development of the market. In the design of system, it should highlight the systemic and integrity. In the product development, it should emphasize innovation and generality.

3.2 *To establish a professional management team and specialized management institutions*

The grasp and development of the talent is the key to the success of an enterprise. In the final analysis, modern enterprise competition is the competition to the talent. Human capital investment is one of the biggest investment. Banks and enterprises have to rely on people to do things to develop. Take the bank of Chongqing for example, since it currently lacks the professional talents of logistics financial business, it should timely, vigorously introduce and cultivate relevant professional talents and retain talents. To the personnel engaged in logistics financial business, they should be more targeted at training. They should not only strengthen financial business knowledge training but also put more attention to teaching and popularizing advanced logistics knowledge and management experience at home and abroad. All these are of great benefit to the development of logistics finance business.

City businesses need to set up specialized management institutions managing the logistics finance business. Only in this way can they carry on the fine operation to improve efficiency and reduce risk. At the same time they should cultivate professional managers who manage the logistics financial business of systematic of banks overall. The organization management structure of the logistics financial business should not only draw lessons from foreign advanced experience, but also accord with China's national conditions. Experience from abroad is that specialized oriented organization structure model is the development direction of the future. Due to the inherent advantages of the line of divisional system itself, foreign banks generally use this pattern. But, most of domestic banks do not have or not reached the requirements of operation departmen. Therefore, taking several walk way to gradually realize this transformation is probably the more stable strategy.

From the point of domestic working, the mode of Guangzhou branch of Shenzhen development bank is worth using for reference. In terms of organizational structure, the branch explores the management mode of platform. And the operation suitable to be centralized, processes and link of requiring specialized skills and the characteristic of scale economy are incorporated into a professional platform. The branch also sets up the energy department, clearing center, ticket center, cargo center, Chu Zhang center and other departments, which embody the flatten management and specialization requirements of operation. In terms of product distribution, the branch explores the characteristic management path of specialization It encourages sub-branch to choose the industry as their main aim according to their own characteristics and resource conditions. Therefore, it can improve the efficiency of using marketing resources and risk control ability. At present, there are eight subsidiaries respectively on energy, automotive, steel, coal, grain and other industry as the main direction to build logistics financial business professional service brand of the bank in the different industries. The platform of architecture and management specialization reflects the interactive requirements. Sub-branches feedback customers' need and market information to branch. And branches provide business guidelines and marketing support, centralized operation and unified standard to sub-branch. In addition, this model will help the branch widen the access of information industry and improve the study ability of industry information. Thus it can provide market information, financial consulting and other value-added services.

3.3 *To reinvent internal management system*

If city business banks want to make the development of the logistics financial business to a new step and become one of their own unique core competitiveness, they have to reengineer process. Process reengineering must reconstruct system at first. Process is dependent on the system, and the system is the guarantee of process to be achieved. Or the process is the expression of the system. In the modern enterprise, system is just likes a "script", and the process is like a "show". Putting the system into the process is just like to put "script" into "TV series". Process reengineering must rely on the system reconstruction. The system of the past was sketchy, management have repetition and conflicts, and even blank. The present system should be meticulous, operable, and fine.

As for the logistics financial service system construction of the city business, developing a programmatic document for the logistics financial business is urgent. According to the master plan, they can also make some specific procedures or rules to form a special system of logistics finance business.

3.4 *To reasonably select the product*

Since market risk exists in the value and the choice of the collateral in the process of selecting the collateral, we must choose qualified popular items. For example, the ownership is clear and complete; Intangible loss is small, not easy to go bad, easy to storage and keep; The fluctuations of the market price is small, relatively stable. Suitable range is large and easy to become cash. Liquidity is better. Specifications are clear and easy to measure. Product quality is qualified and conforms with the relevant standards of the state and so on. It can form varieties list for your reference, and regularly or irregularly revise according to the changing of market situation. Banks can also reduce the risk of the pledge from market value fluctuations by controlling the length of the loan term and the proportion of pledge loan, establishing risk margin method, etc. The pledge selection is vital for risk control of logistics finance business. The choice of the pledge in logistics finance business should consider its liquidate ability and sales ability. You can use the analytic hierarchy process (AHP) that the suppliers often use to choice the quantitative and qualitative analysis. In order to avoid risk better, in view of its professional, if you can introduce the third party who is the pledge appraisal institution with high reputation and good qualification as a partner, and assess the pledge risk and collateral value by these professionals, this will have a role to reduce risk for the development of the logistics financial business of banks.

3.5 *To strengthen the risk management and control tables*

3.5.1 *Manage and control cash flow*

Logistics financial business has the characteristics of self-liquidating, good management and control of cash flow. It is an important way to ensure that the source of repayment. Cash flow management and control refers to the process of using the standard model, designing a suitable product portfolio, signing a binding commerce clause and using financial statements and other measures by the credit bank to track, monitor and manage the flow path of the credit funds after flowing into the enterprise, including starting point, quantity of flow, direction of flow and cycle time, etc. This makes it turnover cycle, value-added return to control cash flow. Also it enhances the predictability, stability and reliability of the source of repayment and ensure enough cash flow to repay debt principal and interest which are maturing.

3.5.2 *Arrange the overall credit reasonably*

Industrial chain as a whole credit is different from a single customer credit, which is on the basis of the banks' consideration in full on a number of customers' financing needs. They are in the industry analysis in-depth and evaluation for the characteristics of the industrial chain and overall ability to resist risks, give credit for industrial chain as a whole and assign to each link of the customers to use. Construction of credit as a whole includes the following four steps; the clear relationship of trade, evaluation to the economic strength for supporting credit trade background, adequate credit arrangements, constraint from risk pricing and terms. Based on the differences of the main risks and benefits range, they can give corresponding risk pricing and business terms.

3.5.3 *Control node of risk*

① The risk control of warehousing partners. For promoting the healthy and orderly development of logistics financial business and standardizing the regulation operation, preventing regulatory risks, city business banks should strengthen warehouse partners'election and management and guarantee to conduct effectively control of confrontation. They can make the management method of warehouse logistics financial institutions in terms of warehousing agency qualifications, cooperation mode, exit, etc.

② The borrower's risk control. In logistics financial business, borrowers repay bank loans with product sales income. And they may not have other material assets on the balance sheet and an independent ability to repay. This is the characteristic of the business. We accept this kind of characteristic. But we should have corresponding request for borrowers' qualifications. For example, the borrower has prominent main business. Operating cash flow is big. Sales channel is unobstructed. He has no bad credit records, etc.

③ The risk control of the downstream enterprise. In industrial products and services, the borrowers' ownstream enterprise qualification must be within the scope of risk control, because the logistics distribution is the most important. Only when the bank can accurately grasp the enterprise goods flow in the transformation process of "cargo – money – cargo", can it effectively judge the possibility of a loan collection.

④ The risk control goods. Logistics financial business requires self-liquidating. which must use these goods that have relatively predictable market risk and strong liquidity. They are easy to liquidate to do pledge goods in order to enhance the second source of repayment. So, they must choose goods reasonably and strengthen the risk control of goods. The goods should have the following characteristics. First, their value is stable and relatively transparent and relatively easy to determine. Secondly, their market demand is large. Thirdly, liquidity is better. Fourthly, quality is stable. The last one, they are easy to store.

⑤ The risk control of chain. In logistics financial business, logistics and information flow are the basic variable to trigger banks to finance and recycle industry chain member enterprises. So in order to realize the orderly cohesion between the logistics, information flow and cash flow, ensure the efficiency and safety of logistics financial business operation, we should establish a cooperation alliance between e-commerce companies and logistics companies and commercial banks. The most direct and regular starting point of logistics financial business development is the core enterprise, which is half way of the international banking practices. Domestic banks have formed a "N alongside 1" and "1 drags N" patterns in order to control the risk of the whole industry chain.

REFERENCES

Jinhui Liu. Several thinking about the commercial Banks to develop logistics financial business [J]. Zhe Jiang finance, 2008(9).

Kai Zhang, Qianli Dong. The financial service innovation oflogistics bank removes the financing obstacles of small and medium-sized enterprise [J]. The theory and practice of finance and economics, 2008(1).

Qin, Zhou. The present situation and tendency of the development of logistics finance in our country [J]. Economic & Trade Update, 2012(5).

Quan Wan. Discussing logistics financial service mode – the third party logistics enterprise's service innovation [J]. Modern service industry, 2009(7).

Industrial Engineering and Management Science – Lee (Ed)
 ISBN 978-1-138-02647-6

Performance evaluation on agricultural convergence development: Taking Beijing Agri-Science-City as an example

Juan Liu, Jiang Zhao, JunFeng Zhang & Jing Gong
Institute of Agricultural Information on Science and Technology, Beijing Academy of Agriculture and Forestry Sciences, Beijing, China

ABSTRACT: Agricultural convergence is a research hot-spot issue. Adopting AHP and Delphi, this paper constructed an agricultural convergence performance evaluation model. Then an empirical study was carried out by taking Beijing Agri-Science-City as an example. Results showed Beijing Agri-Science-City had effectively promoted agricultural convergence development, by the performance evaluation index of 2010–2012 being 0.67, 1 and 1.47 respectively. In addition, based on the analysis of developmental experience and shortcomings of agricultural convergence development in Beijing Agri-Science-City, provided scientific basis for promoting agricultural development.

1 INTRODUCTION

As a revolutionary form of industrial innovation, industrial convergence extended gradually from service industry to the manufacturing and agriculture. It has promoted service economy effectively[1]. The theory and the practice have proved that agricultural convergence development can promote multi-functional agriculture, agricultural chain extension, and agricultural value-added space. So it has been a powerful driving force for agricultural industrial upgrading[2].

Agricultural convergence development had established its advanced theoretical system and development mode at home and abroad[3–5]. But performance evaluation on agricultural convergence development has not been reported. It followed that agricultural convergence is a comprehensive, complex engineering, and the performance evaluation is a new research field.

In 2010 August, the Ministry of science and technology, the Ministry of agriculture and Beijing City Hall jointly launched the Beijing national modern agricultural science and Technology (hereinafter referred to as the Beijing Agri-Science-City). Since the project started, Beijing Agri-Science-City has been practicing agricultural convergence. Taking it as an example to study the performance evaluation can not only enrich the theoretical system, but also guide the practice, having important significance. This paper constructed a performance evaluation model, adopted AHP and Delphi to calculate the weight of evaluation indicators. Then had an empirical study on the convergence development evaluation of Beijing Agri-Science-City from 2010–2012.

2 CONCEPT DEFINITION AND EVALUATION FRAMEWORK

2.1 *Concept definition*

In applied economics field, performance is the general term for the benefit, efficiency and effectiveness, including behavior process and behavior results[6]. Based on the conception, agricultural convergence evaluation included two aspects: drive behavior and development results on agricultural convergence. The definition of performance evaluation on agricultural convergence development is that: according to certain standards and scientific assessment program and by scientific assessment method, regular and irregular assessment has been taken on various kinds of behavior and results about agricultural convergence.

2.2 *Evaluation framework*

Performance evaluation can make qualitative judgments about the capability and efficiency of agricultural convergence development. The evaluation can also find the problems in practices, and provide scientific basis for resources allocation to promote the sustainably development of the industry. Evaluation index system of agricultural convergence development included process index and result index.

From the behavior process, the drive included four broad categories, and they are technical support ability, mechanism guarantee ability, service guide ability and cooperation advance ability[7]. On the other side, from the behavior results, agricultural convergence can contribute to agricultural multi-functionality, centralization of agricultural productive factors, technology innovation of interdisciplinary science, and the

Table 1. Random consistent index coefficient values of comparison matrix.

Order	1	2	3	4	5	6	7	8	9	10	11	12	13	14	15
RI	0	0	0.58	0.90	1.12	1.24	1.32	1.41	1.45	1.49	1.51	1.54	1.56	1.58	1.59

Table 2. The index system and weights of performance evaluation on agricultural convergence development.

Objet layer	Element layer	Criterion layer	Index layer
Performance of agricultural convergence development (A)	Mechanism to promote agricultural convergence (B1) 0.4270	Mechanism of policy linkage (C11) 0.2940	Social capital quantity in Agri-Science-City 0.1866
			Documents distribution jointly 0.1073
		Mechanism of social cognitive (C12) 0.1330	Policy awareness 0.0325
			Research papers number 0.0587
			Number of agricultural tourists 0.0418
	Technical to support convergence (B2) 0.1679	Technology innovation of interdisciplinary science (C21) 0.1078	Number of new varieties in Agri-Science-City 0.0154
			Municipal science and technology progress award of agricultural intersection innovation 0.0347
			Significant innovation project in Agri-Science-City 0.0577
		Transformation of technology achievements (C22) 0.0601	Seed contract transactions in Agri-Science-City 0.0212
			Agricultural technology transactions in Agri-Science-City 0.0243
			Number of trusteeship achievements in Agri-Science-City 0.0147
	Services to guide convergence (B3) 0.2104	Service functional diversity (C31) 0.1313	Operating income of agricultural tourism 0.0490
			Agricultural ecological service value 0.0254
			Number of enterprise in Agri-Science-City 0.0247
			Number of agricultural exhibition in Agri-Science-City 0.0322
		Services elements pluralism (C32) 0.079	Number of agricultural fund in Agri-Science-City 0.0257
			Number of public companies in Agri-Science-City 0.0240
			Information content of Agri-Science-City 0.0174
			Number of net science park0.0119
	Cooperation to advance convergence (B4) 0.1947	Productive factors mobility (C41) 0.1104	Number of international organizations in Agri-Science-City 0.0492
			Land circulation scale in Agri-Science-City 0.0147
			Ratio of agricultural workers 0.0254
			Number of industry alliance 0.0211
		Enhance agricultural competition (C42) 0.0843	Contribution rate of scientific and technological progress 0.0215
			Energy consumption per unit GDP 0.0234
			Agricultural labor productivity 0.0200
			Planting output value per area 0.0196

improvement of agricultural competition. According to the analysis results, the paper established the frame of performance evaluation system.

3 PERFORMANCE EVALUATION MODEL OF AGRICULTURAL CONVERGENCE

3.1 *Evaluation index system design*

In the process of choosing evaluation index, this paper abided by such principles as systematic, comprehensive, independence, representation, comparability, and openness. Basing on the structure diagram of performance evaluation, the hierarchy structural model of evaluation index was devised. The evaluation index system was divided into the target layer, element layer, criterion layer and index layer. Specifically, the target layer is a single index, and the model consisted of 4 elements, 8 criteria and 27 specific indicators.

3.2 *Evaluation index weight calculation*

Adopting AHP and Delphi, the evaluation index weight was calculated. Firstly, expert assessment form was worked out. Then, 15 experts were invited to rank the importance of index layer using 1–9 scale calibration (value is natural number 1–9 and the reciprocal). Besides, the experts came from Beijing agricultural departments, agricultural scientific research, agricultural colleges and universities, and agricultural science & technology zone. Lastly, the judge matrixes were constructed to verify the consistency according to the results of expert scoring. Random consistent index coefficient values of comparison matrix in Table 1. The consistency of the judgment matrix is acceptable.

By square root method, index weights were calculated according to the expert scoring standard. Counting the average of experts scoring weight of each index, the weight of the index system can be calculated. See table 2.

4 ANALYSIS AND RESULTS

4.1 *Data acquisition and processing*

4.1.1 *Data acquisition*

Some of the data were from statistical source, literature resources and survey results. The others were provided by Beijing Municipal Science and Technology Commission, Beijing City Council, Beijing technology market and other related departments. Among the indexes, only "industrial policy awareness" is a qualitative indicator, and the value was calculated through averaging experts scoring.

4.1.2 *Evaluating standard system*

Generally, industry standard, plan standard, historical standard and experience standard can be regarded as the comparison standard in performance evaluation. But it's difficult to ascertain a fixed evaluation standard for agricultural convergence development of Beijing Agri-Science-City. In this paper, the results of 2011 was seen as the reference value to evaluate performance from 2010–2012.

4.1.3 *Data standardization*

The aim of data standardization was to eliminate influence caused by different dimension among the indexes, and the value range is [0.1]. Specifically, this paper regarded the data of 2011 as base value. The data standardization method is: for the positive index of 2010 or 2012, it divided by the corresponding index 2011. For the negative index of 2010 or 2012, its reciprocal divided by the corresponding index 2011. A scale coefficient evaluation index system of agricultural convergence development had been transformed by data standardization.

4.1.4 *Multi-layer evaluation model*

According to the AHP, a multi-layer evaluation model of agricultural convergence development can be described as the following formulae.

$$A = \sum_{i=1}^{m} B_i W_i = \sum_{i=1}^{m}\sum_{j}^{n} C_{ij} W_{ij} = \sum_{i=1}^{m}\sum_{j}^{n}\sum_{k}^{p} C_{ij}^{k} W_{ij}^{k}$$

A: Composite index of agricultural convergence development:

a) W_i: the weight of element lager index;
b) W_{ij}: the weight of the index in I elements layer and J criterion layer
c) W_{ij}^k: the weight of the index in I elements layer, J criterion layer and K index layer

Table 3. Composite index of agricultural convergence development.

Year	2010	2011	2012
Composite index of agricultural convergence development	0.67	1	1.47

Table 4. Element layer index of agricultural convergence development.

Year	2010	2011	2012
Index of mechanism to promote convergence	0.31	0.43	0.59
Index of technical to support convergence	0.11	0.17	0.18
Index of service to guide convergence	0.12	0.21	0.29
Index of cooperation to	0.13	0.19	0.41

4.2 *Evaluation results*

4.2.1 *Composite index of agricultural convergence development*

According to the calculation, the composite index of agricultural convergence development in 2010–2012 was 0.67,1 and 1.47 respectively. The performance level of 2011 was 48.82% higher than that of 2010. Moreover, the performance level of 2012 was 46.87% higher than that of 2011. The reason of slowdown in growth was that Beijing Agri-Science-City was built in 2010, and that make the performance level of 2011 improved greatly because some data of 2010 was zero. Overall, Beijing Agri-Science-City promoted agricultural convergence development effectively, and the efficiency was still lager room.

4.2.2 *Analysis of the main factors*

Analysis of element layer can find the influence degree of drive factors of agricultural convergence development. See table 4. The biggest influence on agricultural convergence development was mechanism. That reflected that the government departments had attached great importance to Beijing Agri-Science-City. From growth trend, the role of cooperation to agricultural convergence increased from 0.13 to 0.41. The growth is significant that reflect Beijing Agri-Science-City speed up cross-industry. About technology and service, the level of those factors greatly enhanced from 2010 to 2011. But in 2012 the growth was slower. Hence, Beijing Agri-Science-City should strengthen technology innovation and service innovation.

4.2.3 *Experiences and problems in practice*

Analysis the 8 criterion layer indexes can ascertain the experience and lack of agricultural convergence practice. See table 5. Some advantages and weakness can be concluded according to the comparison.

Table 5. Criterion layer index of agricultural convergence development.

Year	2010	2011	2012
Policy linkage mechanism	0.215	0.294	0.418
Social cognitive ability	0.093	0.133	0.172
Multidisciplinary science innovation	0.104	0.108	0.129
Sci-tech achievements transformation	0.011	0.06	0.051
Service multifunction	0.096	0.131	0.172
Service elements pluralism	0.027	0.079	0.118
Production factors mobility	0.047	0.111	0.321
Agricultural competitiveness advance convergence	0.08	0.084	0.088

There were four a little bit bright on agricultural convergence in Agri-Science-City. The first one was that interaction among government departments was a catalyst to promote convergence. The second one was that productive factors flow promoted the optimal allocation of agricultural resources. The third one was that modern service promoted the modern multifunctional agriculture. The last one was that the participation of social forces became the booster of agricultural convergence.

On the other hand, there existed some deficiency in agricultural convergence of Agri-Science-City. The first one was that agricultural competitiveness had large room for improvement. The second one was that transformation for agricultural science and technology achievements must be strengthened. The third one was that interdisciplinary innovation and technological innovation industrialization must be strengthened. The last one was that Agri-Science-City should support high-end services to combine with industry chain extension.

5 CONCLUSION

(1) A performance evaluation index system on agricultural convergence was selected. Performance evaluation on agricultural convergence development was a weak link in research. This paper combined process evaluation with result evaluation, and selected a performance evaluation index system including four drive factors: technical support ability, mechanism guarantee ability, service guide ability and cooperation advance ability

(2) A performance evaluation model on agricultural convergence was built. Adopting AHP and Delphi, the evaluation index weight was distributed to ensure the scientific of this research.

(3) Taking Beijing Agri-Science-City as an example. An empirical performance evaluation was carried out. Results showed Beijing Agri-Science-City had effectively promoted agricultural convergence development in 2010–2012. Meanwhile, the composite index of agricultural convergence development increased year by year. According to the results of element layer and criterion layer, some experience and shortcomings of agricultural convergence development in Beijing Agri-Science-City were found out.

Based on the results, it was surely that there was still a greatly potential in Agri-Science-City on agricultural convergence development. The next step was to solve some questions, such as improving the strength of industry body integrating the resource of service body, promoting the technological innovation of industrialization etc, to improve the sustaining power and radiate ability of Beijing Agri-Science-City.

REFERENCES

Fu Da you, Yuan Yong zhi, RUI Guo qiang. Administrative reform and institutional innovation: analysis of local government reform [M]. ShangHai: Shanghai Sanlian Bookstore, 2004. May:57–59

Li Mei yun. New foreign research progress on industry convergence [J]. Foreign Economies & Management, 2005(12):12–20

Liu Juan, Gong Jing, Zhang Xiao hua. Analysis on Industrial Convergence Practice of Beijing Agri-Science-City [J]. Journal of Beijing University of Agriculture, 2014(1): 20–22

Luo Yi. Industrial convergence theory and economic consequence [J]. Economic & Trade Update, 2007(3):37–38

Ma Jian. Industrial convergence theory research review[J]. Economics Information, 2002(5):78–81

Wang Xin-kun. Industry convergence—New concept of agriculture industrialization [J]. Research of Agricultural Moderniaztion, 2007(5):303–306

Zhang Li yang, Zhang Xi cai. Study on the integration of modern agricultural industry chain in China [J]. Teaching and Research, 2007(10):14–19

Industrial Engineering and Management Science – Lee (Ed)
© 2015 Taylor & Francis Group, London, ISBN 978-1-138-02647-6

Innovation of MBA teaching mode based on constructivism learning theory

Ting Wang
Business School, China University of Political Science and Law, Beijing, China

ABSTRACT: Constructivism learning theory highlights in scenario's effect on learning, the construction of learning significance and the learner-centered collaborative study. Under the guidance of constructivism learning theory, a new teaching mode has a main line focused on problem-oriented learning, scenario setting, collaborative dialogue and significance construction. Actually, it should be formed in MBA education.

Keywords: Constructivism learning theory, MBA teaching, significance construction, scenario

1 INTRODUCTION

As MBA education features in special cultivation objects, it requires directed application of related learning theory to guide the practice process so as to improve learning effects. As a kind of emerging learning theory, constructivism proposes a series of new explanations to learning and teaching process. It also provides new teaching ideas and methods for the in-depth development of MBA education.

2 INSPIRATION ON MBA EDUCATION BY THE BASIC IDEA OF CONSTRUCTIVISM LEARNING THEORY

Based on cognitivism, Constructivism learning theory (CLT) is a unique learning concept developed by Vogotsgy, Piaget, Bruner and etc. The fundamental differences between it and the traditional teacher-centered teaching mode are that it highlights on learner's active construction of knowledge based on existing knowledge and experiences. The mode believes that learning is a process in which learner constantly replenish, enrich and transform existing knowledge and experiences. It is based on the mutual effect between new information and existing knowledge and experience. CLT emphasizes more on the initiative, creativity and applicability. And its inspiration on MBA teaching practice mainly embodies in the following aspects.

2.1 *The subject in teaching process is the learner rather than the teacher*

CLT believes that learning is not a process during which the teacher simply transfers knowledge to learner, but a process during which the learner constructs his/her own knowledge. This construction process stresses on subjectivity and selectivity. As a kind of cognition, learning is a process. In this process, the subject actively selects and constructs knowledge, and its key lies in learner's "study". The teacher is providing information to learner rather than transferring knowledge to learner on MBA class. Knowledge can't be transferred to learner from teacher directly. And each learner is accepting different information through selection based on his/her own experiences. As a result, he/she can build his/her own knowledge system, both including the construction of the significance of new information, and the transform and reorganization of existing knowledge experiences.

2.2 *Teaching method emphasizes on problem-oriented independent learning*

CLT thinks that the purpose of learning is for adapting to the world more effectively, so each learner must specify his/her learning purpose. During MBA teaching, learner must actively participate in the teaching process with questions. As different learners have different education backgrounds and work experiences, so the problems that they expect to solve through study are also different. Therefore, teacher should play the roles both as helper and promoter for their knowledge construction on the class. Besides, he must actively mobilize learners to actively participate in the process of solving their own problems.

2.3 *Teaching design advocates situational collaborative teaching method*

CLT thinks that a sound and interactive situational collaborative learning environment must be created for constructing knowledge structure with learner. In MBA teaching, the process of learning knowledge should no longer rely on teacher's direct impartment. But on that learner realizes the construction of learning significance through cooperative learning with the aids by teachers and study partners under a certain

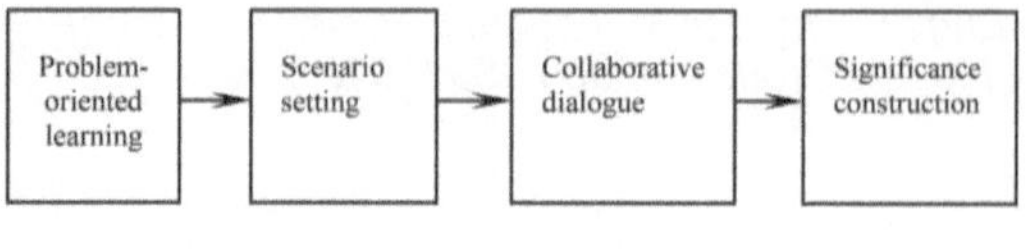

Figure 1. MBA teaching mode based on CLT.

scenario. Namely socio-cultural background, and then to form his/her own knowledge system.

Therefore, the layout of learning environment should not be simplified during the design of MBA teaching. On the contrast, learning should be carried out in a scenario similar to real environment. The learning content should imitate authentic task, so as to promote learner to solve the realistic problems he/she encountered during work.

3 MBA TEACHING MODE INNOVATION BASED ON CLT

CLT stresses on scenario's effect on learning, the construction of learning significance and the learner-centered cooperative learning. Therefore, it advocates that the ideal learning environment must include 4 major elements such as scenario, collaboration, dialogue and significance.

1) The scenario in learning environment must benefit learner to construct the significance of the learned contents;
2) Collaboration includes the cooperation between teachers and learners, as well as the cooperation among learners which should run through the entire teaching process;
3) Dialogue is also a kind of basic cooperation method, requiring learning group members to fully exchange and share their experiences;
4) The construction of significance is the final goal of teaching activity to which everything should be centered on.

According to these key elements, MBA teaching must form a new mode with a main line focused on problem-oriented learning, scenario setting, collaborative dialogue and significance construction (See Figure 1).

3.1 *Problem-oriented learning*

The prominent problems of traditional teaching mode include the lack of learning demands investigation, unclear learning goal and mismatches between learning content and realistic demands. These all show obvious "study and practice contradiction" and cause poor learning effect. CLT highlights in problem-oriented learning. So MBA teaching should specify the learning goal of each lecture. And each learner must extract 3 to 5 realistic problems that they expect to solve during learning process. Then they fill out "learning demand questionnaire". Finally, according to his/her actual work situations before the class, they submit it to the course organizer by combining the learning content. The course organizer should summarize the problems proposed by learners and communicate with the teachers to preset problem scenario when preparing the lessons.

Problem-oriented learning highlights that learner is the subject. Learner must enhance his/her subject consciousness to regard learning as the important way for self development and solving realistic problems both at class and during independent study or cooperative study. This will enable learner to consciously form initiative and creativeness.

3.2 *Scenario setting*

Traditional teaching mode may also require comfortable and humanized learning environment. But the class teaching environment and actual work environment are two totally different scenarios. So it is very difficult for learner to quickly enter the state of solving realistic problems. "To obtain knowledge without context is usually stiff and non-practical", said by educator and philosopher Whitehead. In the design of MBA teaching, to set a scenario which is helpful for learner to construct significance is a very important link.

MBA teaching may effectively combine new information with existing knowledge and experiences to reconstruct to form learner's own knowledge system. It can be done as the way of imitating real working environment through scenario reappearing teaching methods such as case study, role play and team game. MBA teaching may also break the limit of physical space of class. Moreover, the teaching may select such forms as office site teaching, visiting teaching and outdoor teaching to enhance the authenticity, participation and interestingness of learning environment.

3.3 *Collaborative dialogue*

Traditional teaching mode usually limits to the simple mode of "teacher lecturing, learner listening". The mode neglects learner's personalized demands and the interaction between teachers and learners and that among learners. Vygotsky believes that human's advanced psychological skill is developed from the social interactions. Namely, human's advanced psychological activity is sourced from social interactions. The actual practice of collaborative dialogue requires effective design. Teacher is no longer the absolute authority of knowledge. But the helper, cooperator and consultant who timely gives learner with guidance, feedbacks and encouragements.

During the MBA teaching, such collaborative dialogue activities as partner sharing, group discussion and interactions between teachers and learners may be purposely organized according to scenario features, so as to create a harmonious and cooperative learning atmosphere with equal personality. Due to the differences of their industries and posts, the actual scenarios and the problems encountered on the posts of MBA students are not quite the same. Therefore, students may think and analyze independently. Besides, they can express their point of views or doubts by combining the actual problems they encountered during work with the information and analysis mentality provided by the teacher. Then they may make full communication in the form of learning group or class discussion under the guidance of the teacher through brainstorms. They should pay attention to trans-positional consideration and experience sharing discussion. So the entire learning group may jointly complete the construction of significance of the learned knowledge.

3.4 *Construction of significance*

Students in traditional teaching mode usually complain that "I have almost forgotten what I learned after I come back from class". This is because the duck-stuffing teaching mode cannot motivate adult's learning interests. The learner is totally under the state of passively accepting information. The quantity of knowledge obtained by adult mainly depends on learner's ability on constructing the significance of related knowledge. It depends on his/her own experiences, rather than on learner's memory and recitation ability. Therefore, under the guidance of CLT, significance construction is the final goal of entire MBA teaching. Learner must construct the significances of accepted new information and existing knowledge and experiences during the learning process. As a result, they may realize profound understanding on the property and rule of the things reflected from learning content and the internal relation between these things and other things.

Such methods as exploratory teaching method, discovery teaching method and heuristic teaching method should be fully applied in MBA teaching. The methods are used to encourage and require learners to actively collect and analyze related information. Furthermore, they are also used to relate the things reflected from the learning content with their existing knowledge to the largest extent. In addition, thinking such relations carefully is a useful way, so as to promote learner to be the active constructor for learning significance.

Teachers must be the helpers for the construction of learning significance in teaching process. On one hand, they must change traditional teaching mode's silent study atmosphere and motivate learners' learning interests. They also must help learners to form learning motivation through creating scenarios that meet with the demands of teaching content. On the other hand, they must constantly remind the clue for the relation between the newly provided information and existing knowledge and experiences. As a consequence, they can help learners to construct the significance of currently learned knowledge.

4 CLT'S APPLICATION KEYS IN MBA TEACHING

4.1 *Theory acknowledgment is the fundamental guarantee for successful application of CLT*

CLT provocatively subverts the simple talent cultivation way in traditional teaching mode. To enhance the effectiveness and pertinence of learning, it requires meticulous design for the entire process from study demand investigation, selection of lecturing teachers and teaching implementation process to study effect evaluation.

This requires all the involved study subjects including course organizer, teachers and learners to understand and identify the basic ideology of CLT. They also need to cooperate with each other during the implementation process. Then, the new mode can enable the learner to construct the significance of learned knowledge. Finally, the can truly achieve the basic requirements of solving the actual problems he/she encountered in work through MBA teaching.

4.2 *Application of new technology provides conditions for realizing constructivism learning environment*

The successful application of CLT in teaching depends on the establishment of the scenario required by significance construction. Due to the limit of equipments, it is hard for traditional teaching mode to establish a study environment which is good for interaction and realizing sense of experience. With more importance attached to MBA education by colleges and universities in recent years, they have enhanced their investments in the field and equipments for MBA teaching. Especially the full application of computer multimedia technology and network technology makes CLT's many ideologies possible in teaching practice. The professional laboratories were built through full application of multimedia. So the laboratories can create various scenarios through analog simulation methods to enable learners to participate in the study through multiple senses. Furthermore, instant answer and teaching discussion among all computer terminals during the teaching process can be realized by fully utilizing network's remarkable feature of real-time interaction. As a result, it can enable easy collaborative study between teacher and learner and among learners. This kind of network-based collaborative dialogue will effectively motivate learners' strong desire for study, and make them feel the unprecedented study interests under traditional teaching mode. Actually, it is beneficial to perfect and deepen the significance construction of the learning content involved in the discussed topic.

4.3 *Establish evaluation system to verify the teaching effect based on CLT*

Overall process quality control system is also required to be established for the implementation of MBA teaching mode based on CLT. In order to guide all learning subjects to constantly, such system may verify the study effect by setting control points and evaluation indexes at various links. During learning, it can strengthen the study method of constructivism. Therefore, teachers may seriously inspect that if the problems which are expected to be solved by learners are discussed on the class. All the problems should be done through necessary scenario setting and collaborative dialogue during students' cognitive activities in MBA teaching. And teachers may require each learner to fully rethink and write class summary report after the class to answer the questions he/she proposed before the class. This process fully embodies the significance construction of the new information obtained in the class and existing experiences. Meanwhile, it also achieves double study goals of both individual knowledge system establishment and solving realistic problems.

ACKNOWLEDGMENTS

This study was supported by: Humanistic and Social Science Research Fund of Ministry of Education of China (11YJA790153); Program for New Century Excellent Talents in University launched by Ministry of Education of China (NCET-12-0978); the State Scholarship Fund by CSC (201208110408); University Level Humanistic and Social Science Research Program (12ZFG63001); Core Course Research Program for Graduates of Professional Degree (HXKC01) and the Program for Young Innovative Research Team in CUPL.

REFERENCES

Cao Ying: Cultivation of Independent Learning Ability under Constructivism Learning Theory; Journal of Liaoning University of Technology (Natural Science Edition); Issue 1, 2009.

Shu Hongxia, Hu Kunxiu: Basic Theoretical Review on Constructivism Learning Theory; Higher Education Forum; Issue 1, 2009.

Xue Guofeng, Wang Yahui: Analysis on Contemporary Western Constructivism Teaching Theory; Research on Higher Education; Issue 1, 2003.

Industrial Engineering and Management Science – Lee (Ed)
© 2015 Taylor & Francis Group, London, ISBN 978-1-138-02647-6

Author index